User's Guide

Online Journey Through™ Astronomy: Brief Course

Online Journey Through™ Astronomy: Brief Course

Michael Guidry, Margaret Riedinger, and Frank Edward Barnes

Student Companion

written by

Kevin Lee

University of Nebraska–Lincoln

User's Guide

written by

Deborah Gale

BROOKS/COLE

THOMSON LEARNING™

Australia • Canada • Mexico • Singapore • Spain • United Kingdom • United States

BROOKS/COLE
THOMSON LEARNING

Sponsoring Editors: Jennifer Huber, Gary Carlson
Media Editor: Pat Waldo
Assistant Editor: Marie Carigma-Sambilay
Assistant Editor: Samuel Subity
Cover Design: Vernon T. Boes
Cover Illustration: Bob Western

For more information about this or any other Brooks/Cole products, contact:
BROOKS/COLE
511 Forest Lodge Road
Pacific Grove, CA 93950 USA
www.brookscole.com
1-800-423-0563 (Thomson Learning Academic Research Center)

ISBN 0-534-57240-5

Printed in Canada

10 9 8 7 6 5 4 3 2 1

Contents

Introduction

This companion contains numerous exercises to help you master the subject matter contained in *Online Journey Through™ Astronomy*. Although this web site presents a tremendous amount of information in a visually exciting and often interactive manner, only through using these astronomy concepts yourself to solve problems and draw conclusions will you gain a true understanding of them.

Each unit of the workbook is organized in the same manner. The first component is a list of *Chapter Objectives*. These objectives succinctly describe exactly what the authors would like students to learn. The major new vocabulary terms for the chapter are listed in a *Keywords* section. A *Progress Checklist* helps you keep track of which modules you have worked on, and you can check them off as you go. The fourth component is an *Introductory Narrative* describing the content of the chapter in a broad overview. The passage has blanks for you to fill in, most of which can be found in the Keywords section. The blanks in the passage could be filled in as you work through the chapter to reinforce vocabulary and concepts, or afterwards as check of whether or not the major concepts have been retained. This section is designed to allow you an opportunity to use new vocabulary terms very early on in the learning process.

Exercises with various formats will follow depending on what is most suitable for the content of the chapter. Exercises involving graphing of data and drawing the appearance of objects from various geometrical perspectives will occur frequently. Performing simulations using Java™ Applets and analyzing the results will also be a common format. Some sections contain an additional exercise on a particularly important concept (Newton's Laws, for example). The first element of the exercise will typically already be completed to help students get started. Often more insight can be gained by viewing a related interactive component either during an exercise or immediately afterwards, and you will often be directed to do so. The interactive components will be referenced by the code IC followed by the label on the Master List of Animations (for example IC 1.24 would reference the Constellation Viewer).

Each unit concludes with a number of True/False questions. These questions will survey all of the ideas of the chapter and can be used as a final check of your mastery of the material.

Conceptual Maps are included at various intervals in this workbook. These sections tie together the material found in several chapters. The Conceptual Maps differ in appearance from other sections of the book in that they are large drawings that span two facing pages. There is typically an organizational structure such as a timeline or chart with blank labels and descriptions for the you to fill in. The goal behind the Conceptual Maps is to show how the small pieces of knowledge mesh together to form the broad science of astronomy.

The appendix contains solutions to exercises. The answers to all of the Introductory Narratives, True/False Questions, and any exercises of a vocabulary or simulation nature are provided. Asterisks follow the names of any exercises for which solutions or feedback are supplied in the appendix. Solutions to graphical exercises are not provided due to space limitations.

The content, structure, and format of this Student Companion will evolve over time in an effort to better meet the needs of its users. Students and instructors with comments are encouraged to contact the author by e-mail at KLEE6@unl.edu.

Acknowledgments

The author would like to acknowledge the efforts of several people who aided in the development of the student companion. Michael Guidry and Margaret Riedinger participated in early discussions determining the structure and format of the Student Companion, and provided valuable guidance. Margaret Riedinger wrote the chapter objectives and keywords. She also read rough drafts of many of the exercises and suggested improvements.

Five students participated in a seminar class in introductory astronomy at the University of Nebraska–Lincoln and used early drafts of these exercises. Bhee J Arroyo, Scott Jefferson, Jeremy Kalina, Shauna Mullally, and Cody Pearson all provided feedback that ultimately improved this manuscript.

User's Guide

The Student Companion also includes a separate *User's Guide* for *Online Journey™ Through Astronomy*. This guide will help you get started using the *Online Journey* website.

User's Guide

Online Journey Through™ Astronomy: Brief Course

Introduction

Welcome to *Online Journey Through™ Astronomy*—an Internet-based, self-paced educational experience. With its beautiful photographs and illustrations and more than 300 animations and simulations, *Online Journey Through™ Astronomy* will help provide you with a dynamic understanding of our vast universe.

There are four courses, each organized into chapters, which are in turn divided into Interactive Modules, subsections covering specific topics. You can quickly and easily move between chapters and modules with a click of your mouse.

Your course will also link you to other astronomy Internet sites where you can access astronomical newsgroups, photographs, animations, and data to gain a deeper knowledge of specific topics.

This brief guide to *Online Journey Through™ Astronomy* explains how to navigate through the course and how to use its unique features.

I
Getting Started

System Requirements

To use *Online Journey Through™ Astronomy*, you will need a PC or Macintosh with the following configuration:

PC	Macintosh
Windows 95/98/NT 4.0 or later	System 8.0 or later
200+ MHz microprocessor	200+ MHz microprocessor
32 MB RAM	32 MB RAM
1 MB VRAM	1 MB VRA
Monitor with at least 800 x 600 resolution	Monitor with at least 800 x 600 resolution
56 Kbps modem or faster	56 Kbps modem or faster

You will also need the following programs.

- Microsoft® Internet Explorer® 5 or later for Windows, or 4.5 or later for Macintosh; or Netscape 4.7 or later for Windows or Macintosh.
- Java and JavaScript enabled in your browser
- Macromedia Shockwave and Flash Shockwave plug-ins
- Real Media

Note: If you are not sure if you have these programs, click on "Setup and Configuration" after launching Online Journey Through™ Astronomy. *Follow the instructions for downloading the necessary plug-ins.*

Beginning Your Journey

Log on to your Internet Service Provider. Type in the URL given to you by your instructor and press the Enter key to reach the web site for *Online Journey Through™ Astronomy*. You should see the home page shown in Figure 1 on page 4.

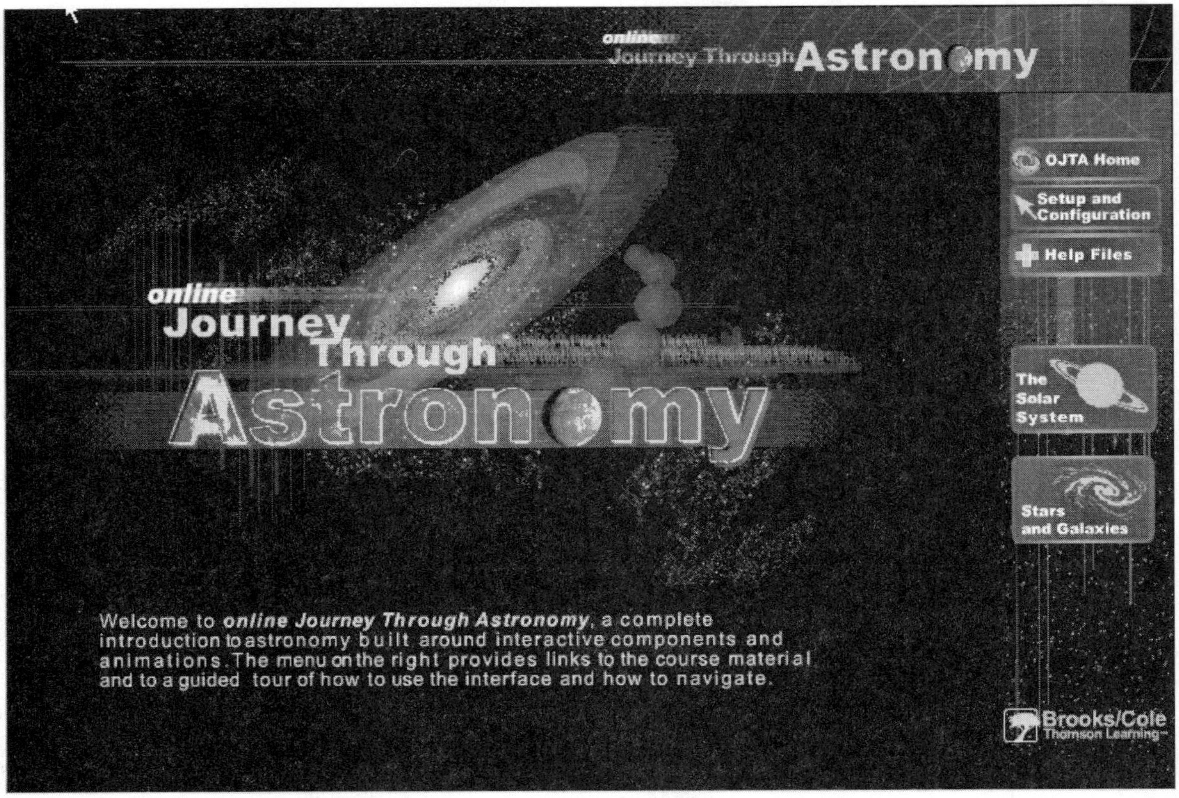

Figure 1. Home page for *Online Journey Through™ Astronomy.*

If you clicked on *Stars and Galaxies*, the table of contents for the course "Stars and Galaxies" will appear (Figure 2).

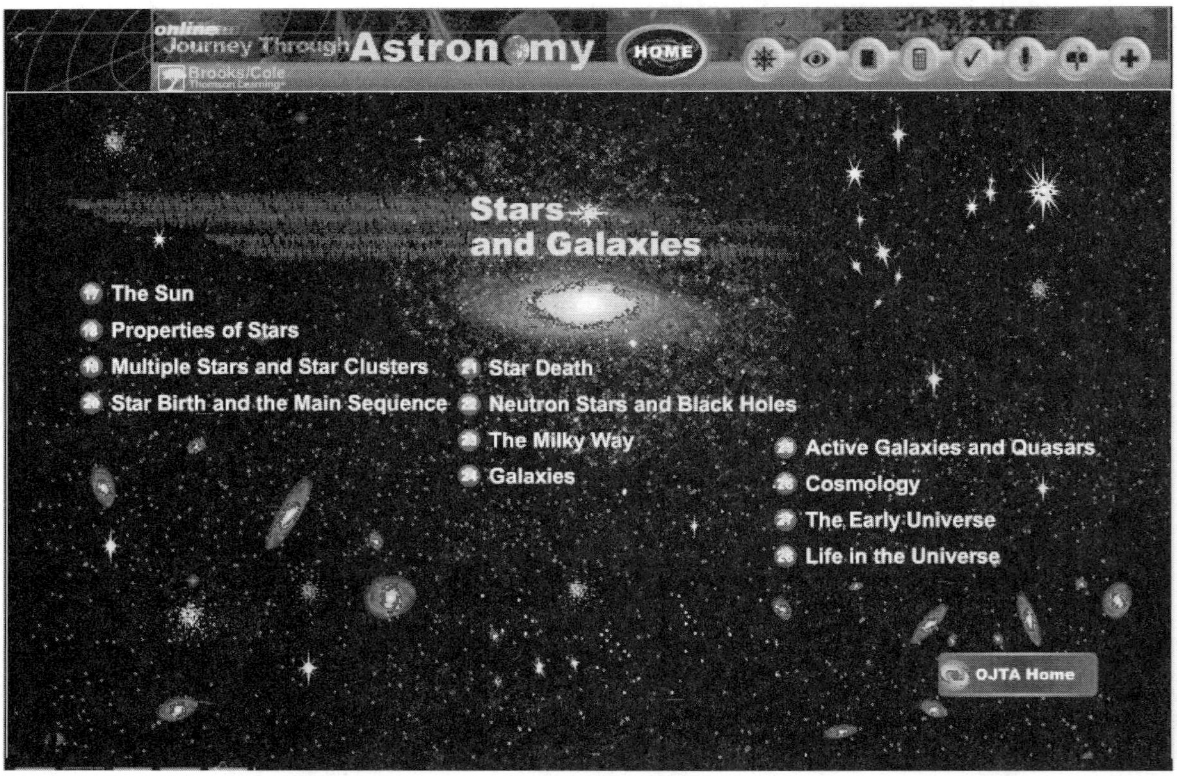

Figure 2. Stars and Galaxies

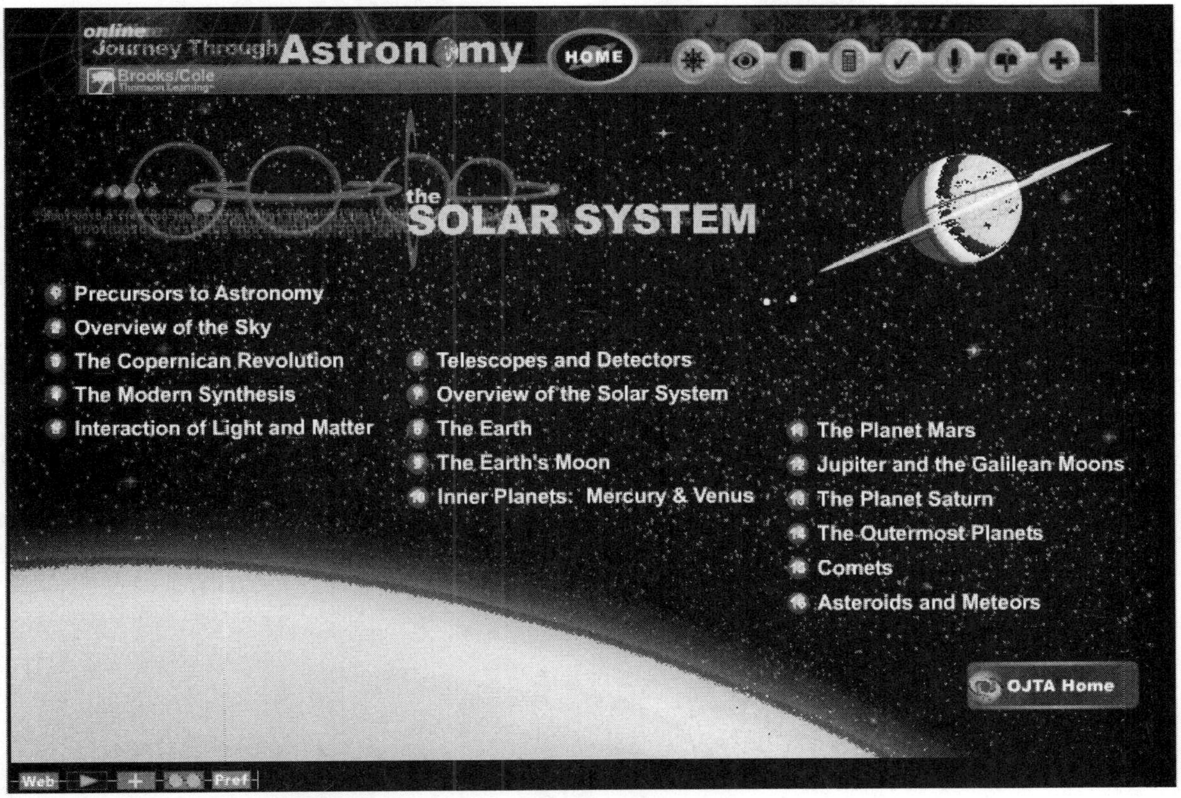

Figure 3. The Solar System

Let's take a quick tour of the menu bar on the right:

Click on this button to make sure that your computer is correctly configured to run the interactive elements in the course.

This button will provide a guided tour of the navigation system and interactive features of Online Journey Through™ Astronomy.

Click on the button that corresponds to the course for which you have enrolled.

If you clicked on *The Solar System*, the table of contents for "The Solar System" will appear as shown in Figure 3.

You can select any chapter in the course by clicking on its title.

At the top right and bottom left of the window are series of buttons called Site Tools. These will link you to various features of the course as well as other astronomy Internet sites. Their functions will be explained in Chapter 3 of the Users' Guide.

In Chapter 2, you will learn how to use the simple and flexible navigation system developed for *Online Journey Through™ Astronomy*.

2
The Navigation System

Online Journey Through™ Astronomy is organized similarly to a textbook, with chapters divided into subsections and special learning tools such as chapter objectives and quizzes. One major difference between a textbook and the online course is that you will click on buttons to move from one chapter to another or to move within chapters. This program has an easy-to-use navigation system that we will explore now.

Course and Chapter Organization

As shown earlier in Figures 2 and 3, your course may contain 12 or more chapters. All chapters have the same basic structure, learning tools, and navigation buttons. Every chapter contains two to eight interactive modules, which are further divided into two to six topics. We will tour a chapter from one of the courses to show how to navigate through the course and within a chapter. Even if you are not taking the course used in the example, just follow along by selecting a chapter in your course.

From the Table of Contents for "The Solar System," click on Chapter 3, "The Copernican Revolution." The following screen will appear.

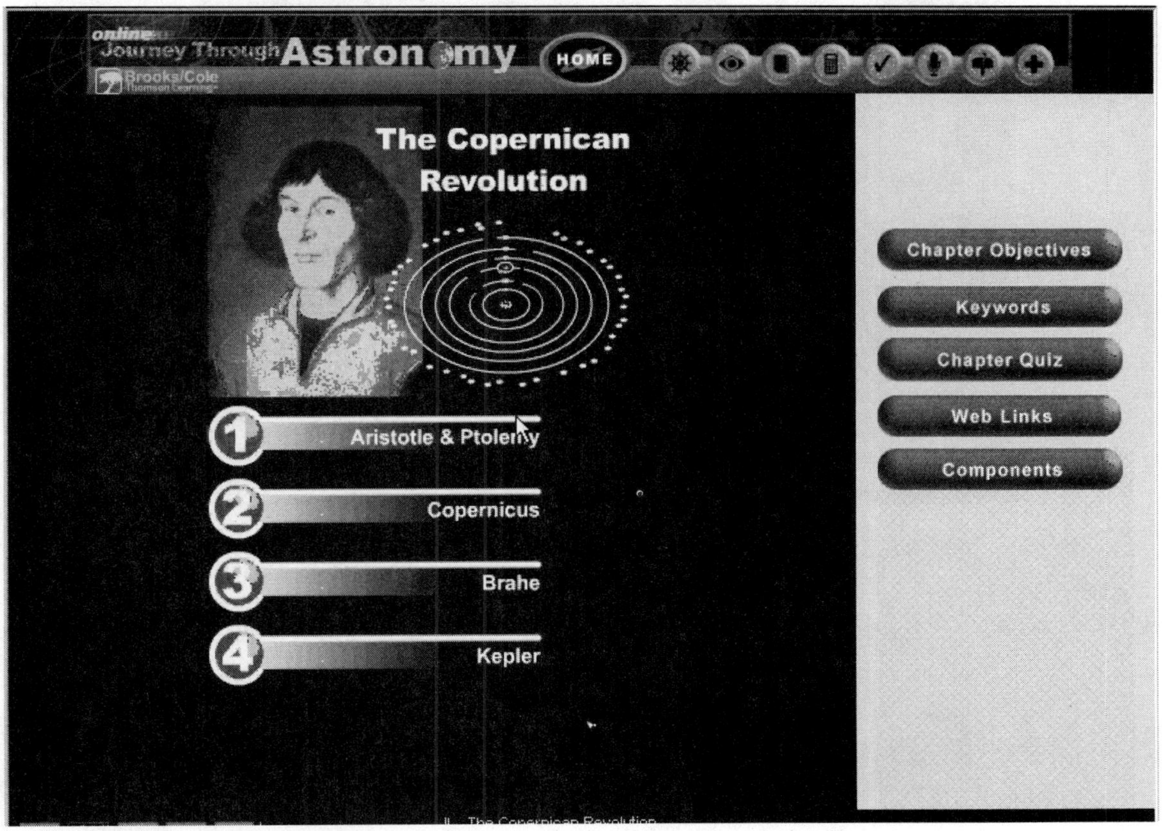

Figure 4. Chapter Home Page.

Every chapter has similar elements: The left side of the screen lists all the Interactive Modules in the chapter. The right side of the screen shows the Chapter Tools, which include Chapter Objectives, Keywords, Chapter Quiz, Web Links, and Components (interactive elements in the chapter).

Click on the button labeled Chapter Objectives.

A window with the learning objectives appears. You can enlarge, shrink, or close this screen by using the standard window buttons or commands on your computer. For example, if you have a PC, you can close this window by clicking on the right corner box (with the X in it).

View the other chapter elements—Keywords, Chapter Quiz, Web Links, and Components—by clicking on those buttons and closing the windows when you are done.

Interactive Modules

Click on the first module, "Aristotle & Ptolemy." You will see the following screen.

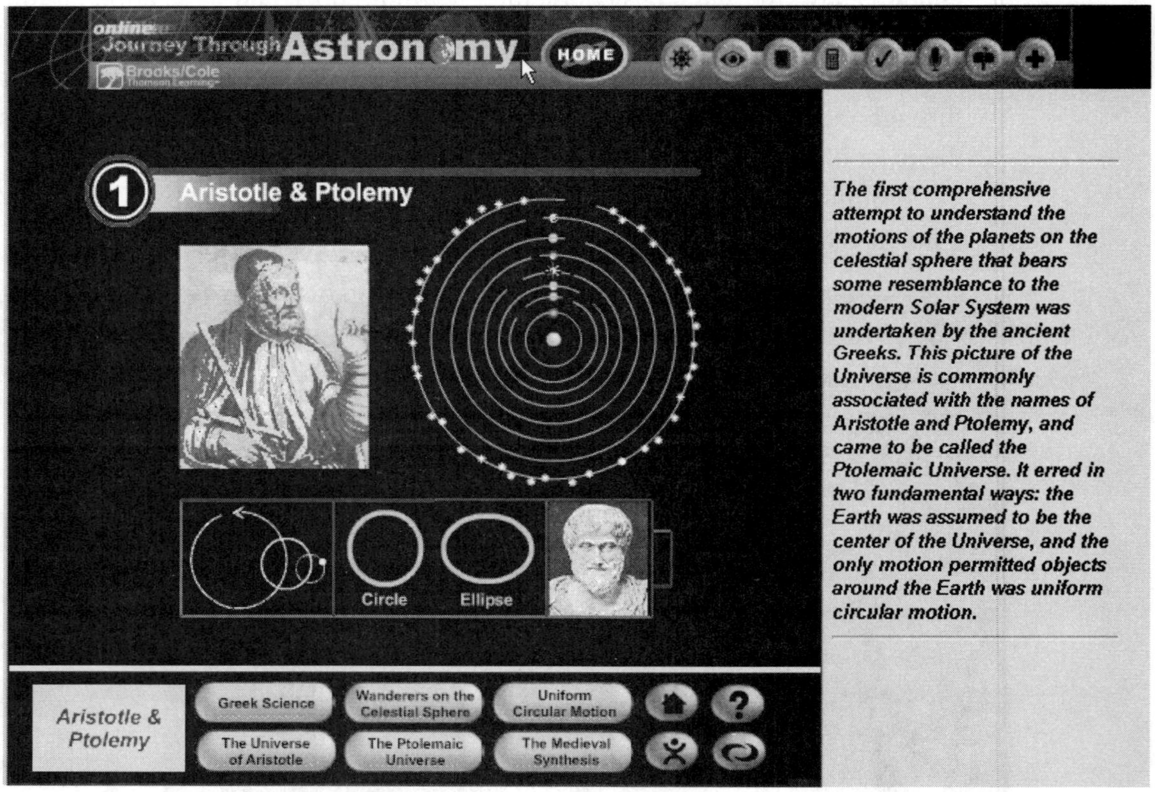

Figure 5. Interactive Module page.

Every Interactive Module in the course is structured similarly to make your journey a smooth one. In the bottom panel are buttons for the topics contained in the module.

Move your mouse over the *Greek Science* button.

Look at the box to the left of the button and you will see a brief summary of that section. Move your mouse over the *The Universe of Aristotle* button, and a brief summary of that section appears. Every module has this feature. Highlight a section button to get a preview of its contents.

Adjacent to the section buttons are four buttons with special icons.

Move your mouse over the button with the question mark icon, and look again at the box on the left to see an explanation. You will see that this button links you to the quiz questions for this chapter.

Clicking on this button returns you to the chapter home page.

This button links to exercises for the chapter.

This allows you to go to relevant astronomy web sites.

On your screen is are the Navigation Bar and Finder Bar. These show you where you are in the course in the format Chapter/Module/Section. The entries in the Navigation Bar at bottom left and Finder Bar to the right are clickable links that load the corresponding pages into the interface.

Navigation Within a Module

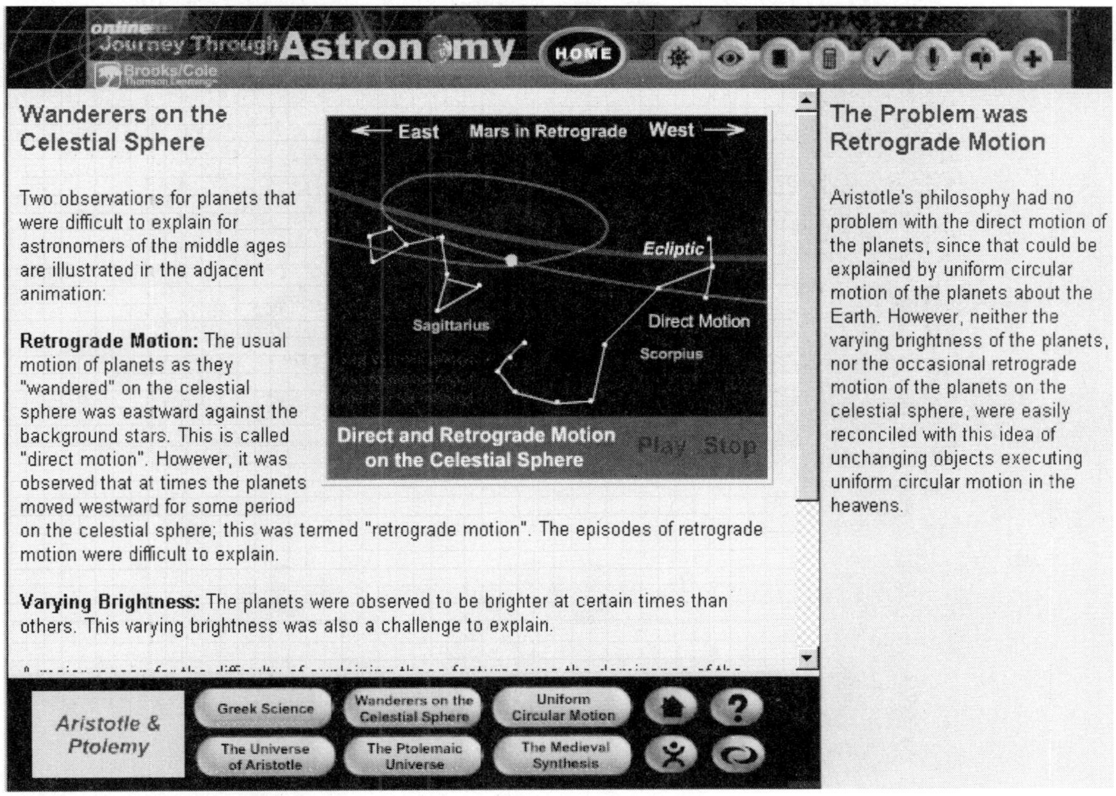

Figure 6. Topic page.

Now click on the topic *Wanderers on the Celestial Sphere*.
You will see the screen in Figure 6.

Notice the scroll bar; it allows you to move down the window on the left. The illustration is animated and is enabled by clicking on a play button. To return to the chapter home page, click on this icon above the navigation bar. Your tour of this chapter is complete. Next you will be introduced to additional tools and features of *Online Journey Through™ Astronomy*.

3
Special Navigation Features

You may have noticed buttons at the top and bottom of the screen. These appear in the same location wherever you travel in the program and have the following functions:

Site Tools 1 (Top)

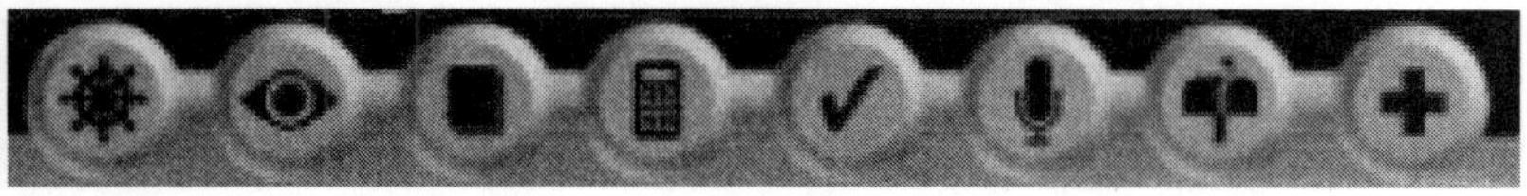

Home: This button returns you to the course table of contents.

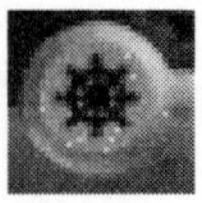

WWW Search: This button allows you to search terms on the Internet.

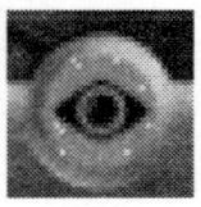

Site Search: This button helps locate specific terms covered in the course.

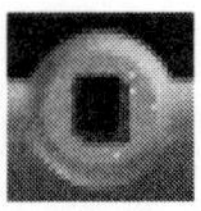

Glossary: Use this button to search for definitions of important terms used in the course.

Calculator: Click on the button and a scientific calculator appears.

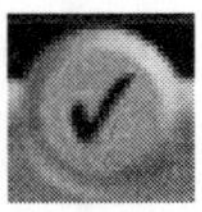

Tutorial Quiz: Use this button to test your understanding of the course material.

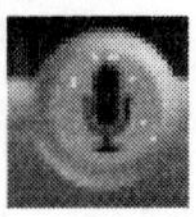

Astro News Group: Click on this button to go to astronomy news groups.

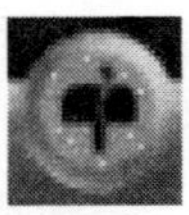

E-mail: This button allows you to send e-mail messages to *Online Journey Through Astronomy*.

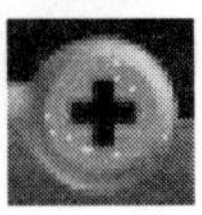

Help Files: If you have questions about using this program, click on this button.

Site Tools 2 (Bottom)

This button links to a set of general web sites for astronomy.

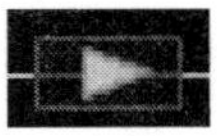

Click this button to view the master list of animations and interactive components in the course.

This provides a list of the most recent additions to the syllabus.

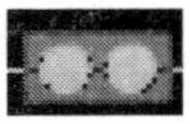

This button lists credits and acknowledgements for *Online Journey Through™ Astronomy*.

You can set preferences for the display view.

Now that you have been introduced to the navigation tools and major features of *Online Journey Through™ Astronomy*, you are ready to begin exploring the universe on your own. Enjoy the experience!

Student Companion

Online Journey Through™ Astronomy: Brief Course

Unit I
Overview of the Sky

Chapter Objectives

Historically, the regular apparent motions of the heavens provided many of the basic ideas, terminology, and impetus to keep careful astronomical observations that would become so important in the development of the science. These cycles of celestial motions serve as cosmic clocks and yearly calendars. In this chapter we will see how the seasonal changes and the daily and hourly changes are all astronomically based. The systems of nomenclature for the stars and constellations and the use of star maps will be introduced. The celestial coordinate system will be explained and used to find positions in the sky. The reason that time kept by the stars (sidereal time) must be different than the solar time (by which we live our lives) will be presented. The different types of calendars will be described and the problems that led to the introduction of more accurate calendars will be outlined. The cause of Earth's seasonal effects will be explained and the solar positions that mark the precise beginnings of these seasons will be discussed. Both ancient and modern ways of classifying the planets in our Solar System will be explained. The basic motions in our sky will be described, and the terminology used to denote these motions will be introduced. The aspects and phases of the planets will be described.

Progress Checklist

1. The Celestial Sphere
- ❑ The Celestial Sphere
- ❑ The Ecliptic
- ❑ The Coordinate System
- ❑ Equinoxes and Solstices
- ❑ Motion on the Celestial Sphere
- ❑ East and West

2. The Constellations
- ❑ Groupings and Asterisms
- ❑ Classical Constellations
- ❑ Modern Constellations
- ❑ Constellation Viewers
- ❑ Star Maps
- ❑ Naming the Stars

3. Aspects and Phases of Planets
- ❑ Classification
- ❑ 7 Planets of the Ancients
- ❑ Stars and Planets
- ❑ Wanderers
- ❑ Inferior Planets
- ❑ Superior Planets

4. Timekeeping
- ❑ Sidereal and Solar Time
- ❑ Sidereal and Solar Days
- ❑ Precession of the Earth's Axis
- ❑ Months and Years
- ❑ Time Zones
- ❑ Calendars

5. The Seasons
- ❑ Northern Hemisphere
- ❑ Southern Hemisphere
- ❑ Lag of Seasons
- ❑ Midnight Sun

6. The Orbit & Phases of the Moon
- ❑ Revolution in Orbit
- ❑ Lunar Phases
- ❑ Rotational Period
- ❑ Tidal Locking

7. Lunar and Solar Eclipses

- ❑ Frequency of Eclipses
- ❑ Geometry Solar Eclipses
- ❑ Types of Solar Eclipses
- ❑ Total Solar Eclipses
- ❑ Eclipse Patterns and Cycles
- ❑ Lunar Eclipses

Keywords

inferior planet
superior planet
Terrestrial planet
Jovian planet
celestial sphere
zenith
celestial meridian
diurnal motion
celestial equator
celestial pole
ecliptic
aspects
phases
superior conjunction
inferior conjunction
elongation
opposition
quadrature
celestial sphere
latitude
longitude
constellation
asterism
zodiac
Orion
Cygnus
Bayer system
Flamsteed system
celestial coordinate system
celestial equator
north celestial pole
south celestial pole
right ascension
declination
vernal equinox
autumnal equinox
summer solstice
winter solstice
sidereal time
solar time
sidereal day
solar day
time zones
Universal Time
lunar calendar
solar calendar
new moon
waxing
waning
quarter moon
gibbous
full moon
perigee
apogee
tidal coupling
solar eclipse
lunar eclipse
umbra
penumbra
total eclipse
partial eclipse
annular eclipse
path of totality
Bailey's beads
diamond ring effect
corona
saros
tides
differential forces
spring tides
neap tides
Gregorian calendar
Julian calendar
seasons
lag of the seasons
maximum insolation
precession
Polaris
precession of the equinoxes
sidereal month
solar month
synodic period
phases

Exercise 1-1: Introductory Narrative

Astronomers talk about the motion of objects in the sky as being movement on the 1) ______ ________________, a monstrous hollow shell that surrounds the entire universe. We denote positions on this surface using the coordinates Right Ascension and 2) __________________. The direction of increasing Right Ascension is referred to as 3) ______________________. The positions of stars in these coordinates changes very slowly over time due to the 4) _____ _________________ of the stars. The coordinates also change slightly due to the motion of the axis of rotation of the Earth. This phenomenon is known as 5) ______________________.

The apparent circular motion of the stars due to the rotation of the Earth is known as 6) ____________________ and doesn't change the coordinates of stars since the coordinate system is fixed in space.

Since the coordinates of stars are fairly stable, patterns in the sky are long-lived. There are 88 official 7) ____________________ that denote groupings of stars. These were important historically for storytelling and astrology, but are still used by astronomers today to uniquely specify a certain area of the sky. Other more modern patterns such as the "Big Dipper" are known as 8) ____________________.

There are many differences between the appearances and motions of planets and stars. Unlike stars, the coordinates of planets on the celestial sphere are constantly changing. In fact, the term planet originally meant 9) ____________________. In ancient times there were thought to be seven planets: Mercury, Venus, Earth, 10) ____________________, Jupiter, Saturn, the 11) ____________________ and the sun. It was especially difficult for ancient astronomers to understand how a planet's brightness could vary and how it could turn around and move backwards for a while, which is known as 12) ____________________. Another difference is that stars could be anywhere, while planets were only found near the 13) ____________________, the yearly path of the sun. In addition, stars twinkle in the sky while planets do not. Attempts to understand these differences led to the beginning of modern astronomy.

Exercise 1-2: Using Right Ascension and Declination

The purpose of this assignment is to gain familiarity with the celestial equatorial coordinate system. In the table below the right ascension and declination are listed for 21 stars in a small region of the sky. Notice that the magnitudes of the stars are given as well. You should first create a legend which details the symbols you plan to use for stars of each magnitude range. (Typically, brighter objects are denoted by larger or darker symbols.) You should then graph the location of each of these stars on the section of the celestial sphere map that follows. Look for information on the **APOD** internet site on M42 that will help you draw the object as realistically as possible.

Note that a title box appears at the top of the star map. Can you use the *Interactive Constellation Viewers* to identify this constellation? (These viewers are found in the *Overview of the Sky/Constellation Module*.)

Star Name	Right Ascension	Declination	Magnitude
Rigel	a = 5^h 14'	d = −8E 12'	0.2
Betelguese	a = 5^h 55'	d = +7E 24'	0.5
Bellatrix	a = 5^h 25'	d = +6E 21'	1.6
Saiph	a = 5^h 47'	d = −9E 40'	2.1
Alnilam	a = 5^h 35'	d = −1E 14'	1.7
Mintaka	a = 5^h 31'	d = −0E 18'	2.3
Alnitak	a = 5^h 41'	d = −1E 57'	1.7
M42	a = 5^h 34'	d = −5E 10'	****
τ Orionis	a = 5^h 17'	d = −6E 51'	3.6
μ Orionis	a = 6^h 02'	d = +9E 39'	4.1
χ Orionis	a = 6^h 12'	d = +14E 13'	4.5
υ Orionis	a = 6^h 07'	d = +14E 47'	4.4
χ1 Orionis	a = 5^h 54'	d = +20E 17'	4.4
χ2 Orionis	a = 6^h 04'	d = +20E 12'	4.6
π1 Orionis	a = 4^h 55'	d = +10E 09'	4.6
π2 Orionis	a = 4^h 51'	d = +8E 54'	4.3
π3 Orionis	a = 4^h 50'	d = +6E 58'	3.2
π4 Orionis	a = 4^h 51'	d = +5E 36'	3.7
π5 Orionis	a = 4^h 54'	d = +2E 26'	3.7
π6 Orionis	α = 4^h 58'	δ = +1E 43'	4.5
Meissa	a = 5^h 35'	d = +9E 56'	3.4

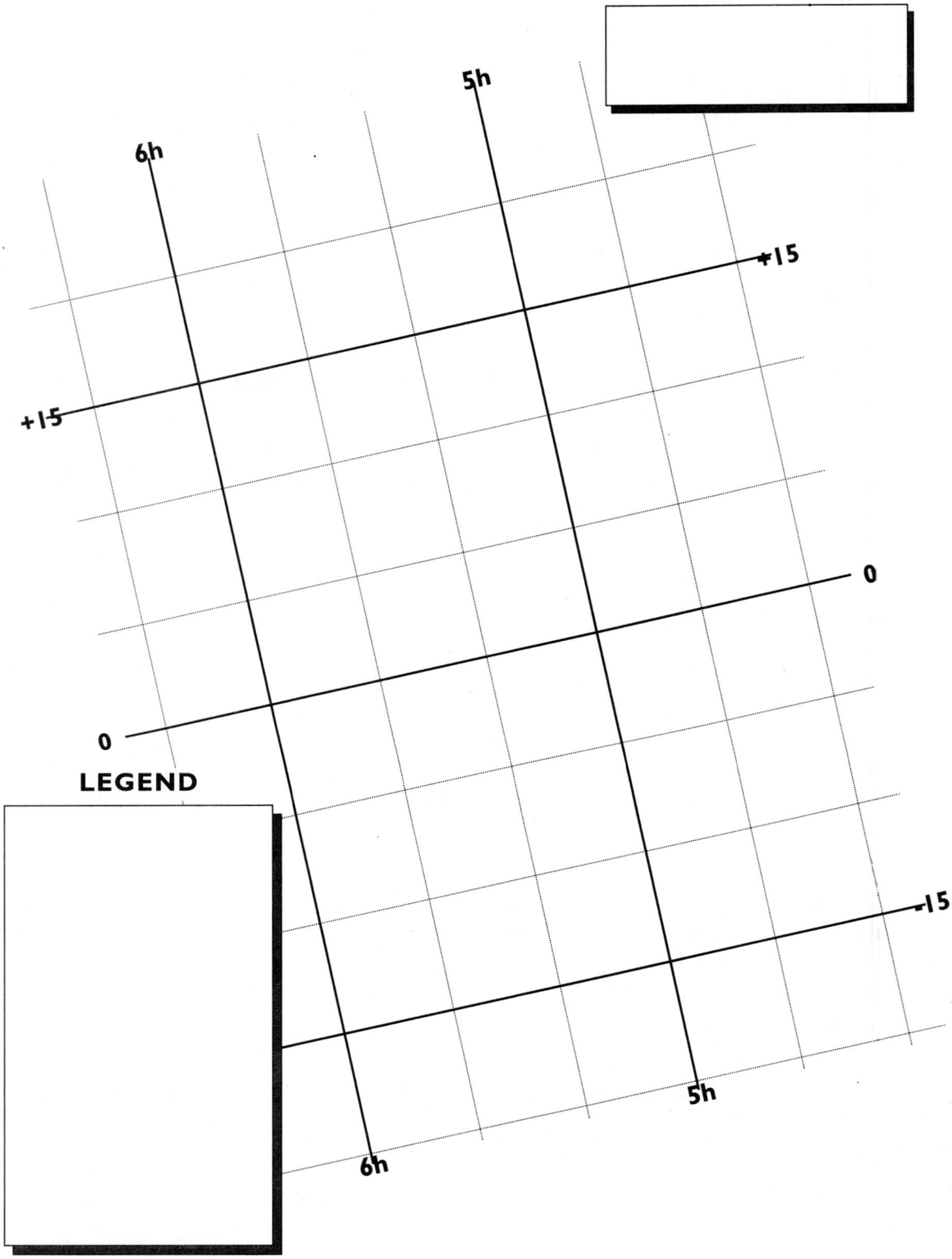

5h
6h
+15
+15
0
0
LEGEND
-15
5h
6h

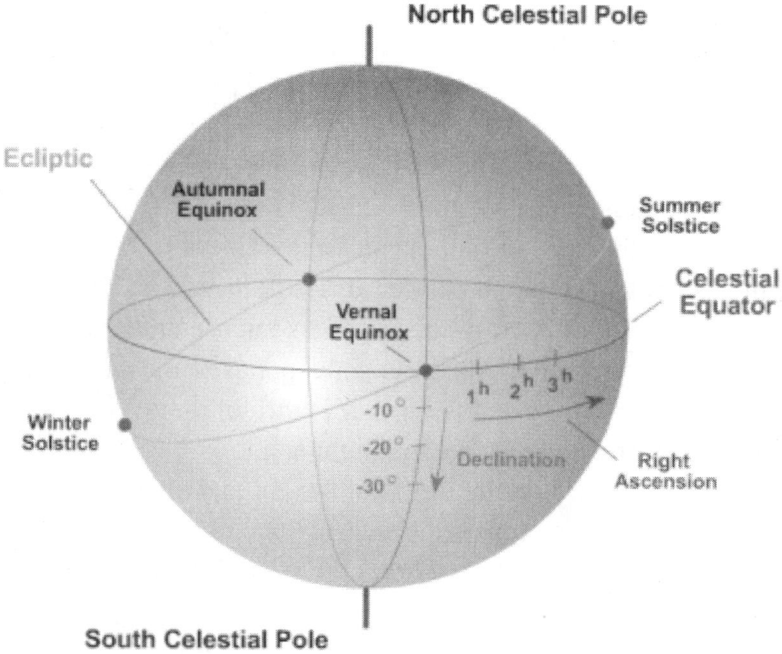

Exercise 1-3: Following the Sun's Path for One Year*

In this exercise we will verbally and graphically follow the sun throughout the year, noting the changes in its position on the ecliptic of the celestial sphere, and where its direct rays hit the Earth. A passage follows, describing the motion of the sun with blanks to fill in. The blank may require a term such as *summer solstice*; *northern* or *southern* to denote a particular hemisphere; or a particular value of latitude or declination. Note the flat map of the celestial sphere on the following page, which shows the ecliptic, and the graphic of the Earth. At several times during the passage the ***Graph*** symbol will appear, directing you to map the current position of the sun and telling you what symbol to use. You should then draw in the indicated symbol on the flat map of the celestial sphere, indicating where the sun is presently located. You should also draw in the indicated symbol on the map of the Earth at the latitude where the direct rays of the sun are hitting at that particular time of year.

One fact to keep in mind that will help you complete this exercise is this: The declination of the sun on the celestial sphere at a particular time is equal to the latitude on the earth where the direct rays of the sun are hitting. You will also find the table of important parallels of latitude useful.

Important Parallels of Latitude	
90° N	North Pole
66.5° N	Arctic Circle
23.5° N	Tropic of Cancer
0°	Equator
23.5° S	Tropic of Capricorn
66.5° S	Antarctic Circle
90° S	South Pole

We will begin our story of the sun's path on March 21, the vernal equinox. The sun is crossing the celestial equator and is moving northward on the celestial sphere. (***Graph*** = "VE") The direct rays of the sun are hitting on the Earth's 1) ______________________. Day and night are 12 hours long all over the entire Earth.

As the days, pass the sun continues to move northward on the celestial sphere. The direct rays of the sun are hitting at progressively higher latitudes. Days are getting longer in the 2) ______________________ hemisphere and shorter in the 3) ______________________ hemisphere.

The sun reaches its most northern point on the celestial sphere on June 21, the 4) ______________________. It is now summer in the 5) ______________________ hemisphere. (***Graph*** = "SS"). On this date the sun has a declination of 6) ______________________ and its direct rays are hitting on the Tropic of Cancer at a latitude of 23.5° N. If one considers progressively more southern latitudes, the rays become less intense until, at 90° away from the Tropic of Cancer, the sun is on the horizon. This corresponds to the Antarctic Circle, where the least direct rays of the sun are hitting. At all latitudes north of the Arctic Circle the sun doesn't set on this day, a phenomenon known as the midnight sun.

The sun now moves southward on the Celestial Sphere and on 7) ______________________, the Autumnal Equinox, the ecliptic crosses the celestial equator. An observer at a latitude of 8) ______________________ can now see sun at their zenith at noon. (***Graph*** = "AE") As the days pass the sun continues to move 9) ______________________ on the celestial sphere. The direct rays of the sun are hitting at progressively more southern latitudes. Days are getting longer in the 10) ______________ hemisphere and shorter in the 11) ____________ hemisphere.

The sun reaches its most southern point on the celestial sphere on 12) ______________, the 13) ______________________. It is now summer in the 14) ______________________ hemisphere. (***Graph*** = "WS"). On this date the sun has a declination of 15) ______________________ and its direct rays are hitting on the Tropic of Capricorn. At all latitudes north of 16) ______________________ the sun doesn't rise on this day. The sun now starts moving northward and the yearly cycle repeats.

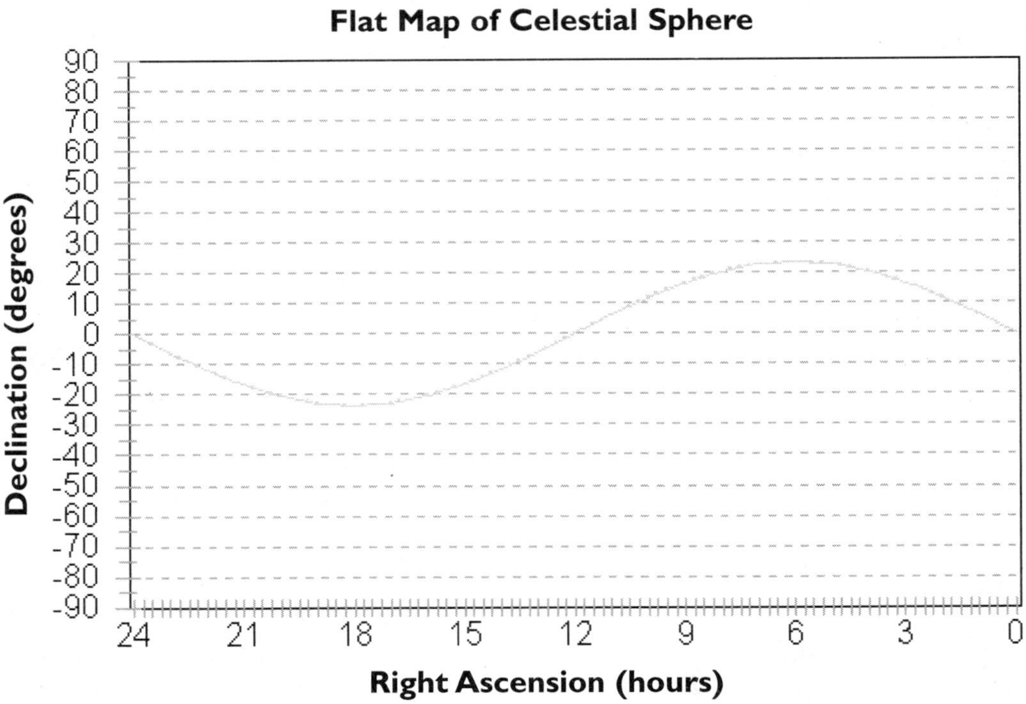

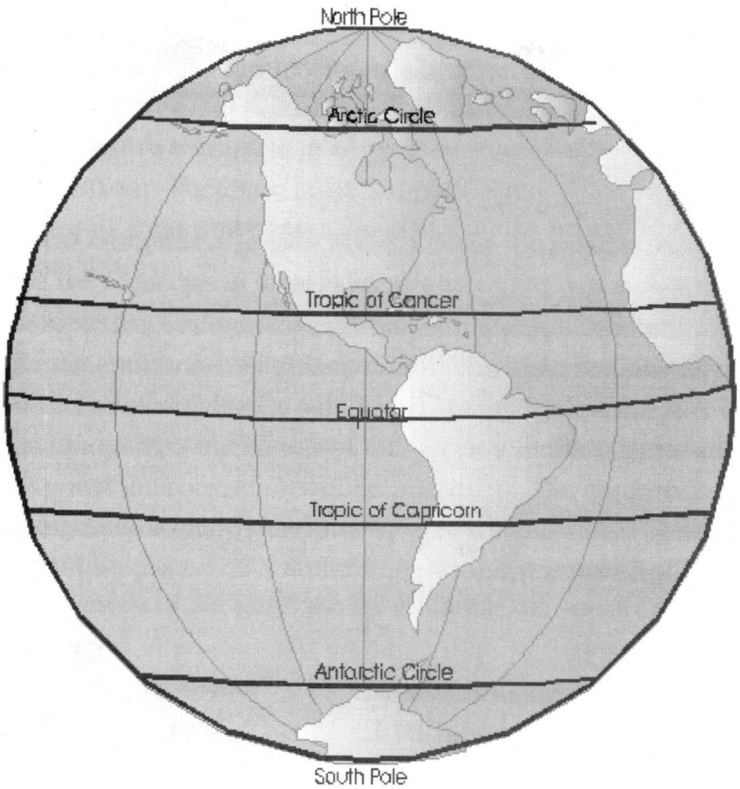

Exercise 1-4: Using Elongation Terminology

In this exercise we will practice with a coordinate known as elongation that is useful for specifying the position of planets. Elongation is defined as the angle between the sun and a planet as viewed from the Earth. It ranges from 0° to 180° and is denoted as either eastern or western, depending on whether the object is east or west of the sun. Special terms such as conjunction, opposition, and quadrature are used to describe several significant values of elongation.

A map of the orbits of the first six planets of our solar system follows on the next page. Our perspective for this map has us looking directly down onto the plane of the solar system. This map is known as a curtate chart. The drawing is not to scale, since it is not possible to effectively draw the orbits of inferior and superior planets with proper scale at the same time. The sun is shown at the center of the solar system and the position of the Earth at a particular time is shown. Draw a line from the Earth to the Sun. This line will form the basis for our elongation coordinate. Elongation is the angle between this line and a line drawn from the Earth to a particular planet. If the planet is to the left of the sun (at a larger right ascension than the sun) the elongation is eastern, as opposed to western to the right of the sun. Note that some values of elongation are ambiguous; Venus at a western elongation of 15° could be at two possible locations on the curtate chart. It is also possible to specify elongation terminology that makes little sense such as "Jupiter at greatest elongation." Greatest elongation is a term that applies only to inferior planets.

Draw in the letter denoting the location of the planet on the curtate chart for each of the following descriptions of elongation. The first elongation specification is completed for you. Use the Interactive Components (**IC 2.11**) and (**IC 2.12**) to check your answers to the rest.

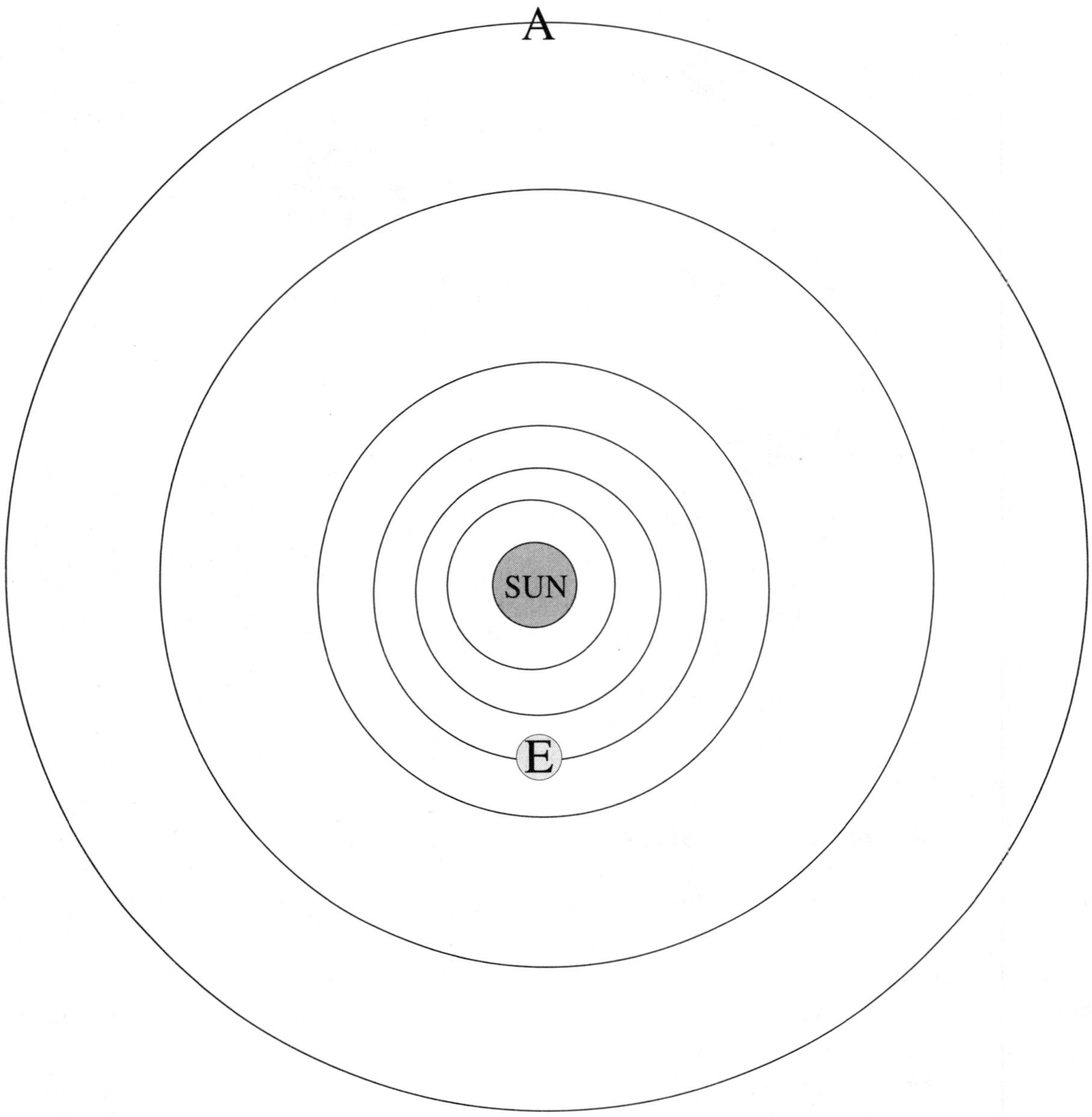

Curtate Chart of the Solar System

A. Saturn at conjunction
B. Mercury at inferior conjunction
C. Mars at western quadrature
D. Jupiter at opposition
E. Venus at greatest eastern elongation
F. Mercury at greatest western elongation
G. Jupiter at eastern quadrature
H. Venus at superior conjunction
I. Saturn at an eastern elongation of 60°
J. Venus at opposition

Exercise 1-5: Rising, Meridian, and Setting Times of Lunar Phases*

In this exercise we will note the rising, meridian, and setting times of the 8 lunar phases. The procedure described here is fairly crude in that it assumes that the moon is always up in the sky for twelve hours. We will only allow the 8 phases and 8 times (3, 6, 9, and 12 A.M. and P.M.) on the Earth. Thus, this method only allows predictions to be made with an accuracy of an hour or so.

The key to doing this is to note the meridian time–the time when a particular phase of the moon is on the observer's celestial meridian. In the diagram below, note that the Earth's time-zones are determined by where sunlight is hitting. Our perspective is looking down from near the North Celestial Pole so that the Earth is spinning counterclockwise. To find the meridian time for a particular phase, we simply need to draw a line from the center of the Earth to the center of the moon for that lunar phase. For example, for the waxing crescent, the line drawn from the center of the Earth to the center of the moon intersects the surface of the Earth in the 3 P.M. timezone. Thus, the waxing crescent crosses the observer's celestial meridian at 3 P.M. To calculate the rising time for this phase we simply subtract 6 hours. The waxing crescent rises at 9 A.M. For the setting time we add 6 hours to the meridian time to get 9 p.m.

Rising, Meridian, and Setting Time Questions

1. The meridian time of the third quarter moon is ____________.
2. The meridian time of the waxing gibbous is ____________.
3. The meridian time of the ____________ is 6 P.M.
4. The meridian time of the ____________ is 9 A.M.
5. The rising time of the waning gibbous is ____________.
6. The rising time of the full moon is ____________.
7. The ____________ moon rises at midnight.
8. The ____________ moon rises at 9 A.M.
9. The setting time of the third quarter is ____________.
10. The setting time of the waning crescent is ____________.
11. The ________________ moon sets at 3 A.M.
12. The ________________ moon sets at midnight.

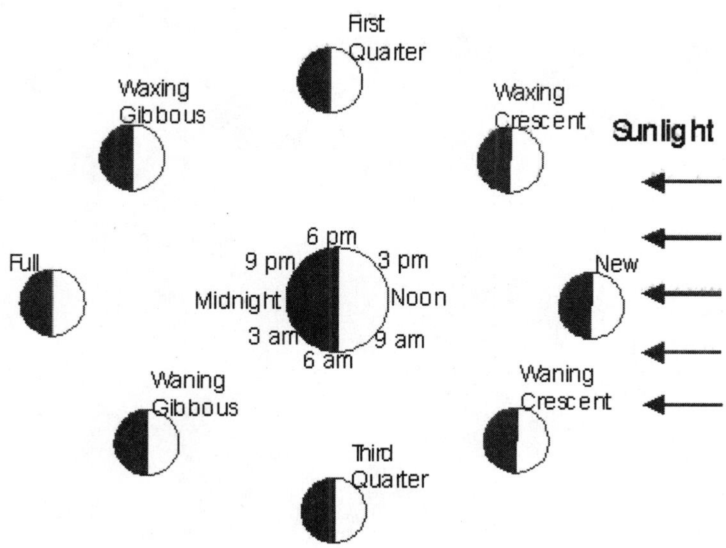

Exercise 1-6: True/False Questions

T / F 1. The point directly above our heads on the celestial sphere is called the nadir.

T / F 2. The coordinate on the celestial sphere analogous to longitude is declination.

T / F 3. The Earth's axis of rotation is inclined by 23.5° relative to the perpendicular (normal) to its orbital plane.

T / F 4. The point of the ecliptic where the sun is most south of the celestial equator is called the summer solstice.

T / F 5. All of the stars in a certain constellation must be located nearby each other in a small region of space.

T / F 6. The Big Dipper is a well-known constellation.

T / F 7. The sun and moon are members of the seven planets of the ancients.

T / F 8. The eastward motion of a planet on the celestial sphere is known as retrograde motion.

T / F 9. Venus is an inferior planet.

T / F 10. The moon's synodic month is 27.3 days long.

T / F 11. Since the moon always keeps the same face pointed towards the Earth, it obviously doesn't rotate.

T / F 12. Lunar eclipses occur during the time of full moon.

T / F 13. Annular solar eclipses occur when the moon is near apogee.

T / F 14. If a total solar eclipse occurs somewhere on the Earth on January 1, the soonest another total solar eclipse could occur is around July 1 of that same year.

Unit 2
The Development of Modern Astronomy

Chapter Objectives

Understanding the important philosophical and scientific changes initiated by Copernicus, with his sun-centered view of our Solar System, is the first step in the development of the modern worldview. This chapter will show how the heliocentric system gives an easily pictured explanation of the retrograde motion of the planets and their varying brightness in our sky. We will also give credit to Aristarchus of Samos, who in 200 B.C., first proposed such a sun-centered idea. The contributions of others who followed after Copernicus will also be described. The observations of Tycho Brahe and their important role in Johannes Kepler's work will be outlined. Kepler's conclusions about planetary orbits, summed up in his Three Laws of Planetary Motion, were essential in our growing knowledge concerning our Solar System.

Galileo's historic telescopic observations and his lesser known contributions to Newton's understanding of dynamics will be explained in this unit. Isaac Newton's culminating work on dynamics (Newton's Three Laws of Motion) and his Universal Law of Gravitation and their importance will be explained. Albert Einstein's more modern theory concerning gravity will bring us to our present-day understanding of this most basic interaction of nature.

Progress Checklist / Precursors to Astronomy

Check off each section after you have studied it online and in this Student Companion.

2. Archaeoastronomy

- ❑ Stonehenge
- ❑ Maya Astronomy
- ❑ Chinese Astronomy
- ❑ North American Indian Astronomy
- ❑ Classical Greek Astronomy
- ❑ Islamic Astronomy

Progress Checklist / The Copernican Revolution

1. Aristotle and Ptolemy

- ❑ Greek Science
- ❑ Wanderers on the Celestial Sphere
- ❑ Uniform Circular Motion
- ❑ The Universe of Aristotle
- ❑ The Ptolemaic Universe
- ❑ The Medieval Synthesis

2. Copernicus

- ❑ Sun-Centered Solar System
- ❑ Retrograde Motion
- ❑ The Copernican Revolution
- ❑ Aristarchus of Samos

(*continued*)

3. Brahe
- ❑ The Life and Times of Tycho Brahe
- ❑ Brahe's Instruments
- ❑ Tycho's Supernova
- ❑ Tycho's Comet
- ❑ Brahe and Stellar Parallax
- ❑ Brahe's Observations

4. Kepler
- ❑ The Life of Kepler
- ❑ Kepler and the Elliptical Orbits
- ❑ Some Properties of Ellipses
- ❑ Kepler's First Law
- ❑ Kepler's Second Law
- ❑ Kepler's Third Law

Progress Checklist / The Modern Synthesis

1. Galileo
- ❑ The Life of Galileo
- ❑ Galileo and the Telescope
- ❑ New Telescopic Observations
- ❑ Leaning Tower
- ❑ Inertia
- ❑ Dynamics of Galileo and Aristotle

2. Newton
- ❑ Sir Isaac Newton
- ❑ Vectors
- ❑ Velocity and Acceleration
- ❑ Newton's First Law
- ❑ Newton's Second Law
- ❑ Newton's Third Law

3. Universal Gravitation
- ❑ The Apple
- ❑ The Orbital Cannon
- ❑ Law of Gravitation
- ❑ Center of Mass
- ❑ Kepler's Third Law
- ❑ Weight and Mass

4. Synthesis
- ❑ Acceleration on Keplerian Orbits
- ❑ Newton's Laws and Kepler's Laws
- ❑ Conic Sections
- ❑ Gravitational Orbits

5. Perturbations
- ❑ 2-Body Approximation
- ❑ Perturbations
- ❑ Discovery of Neptune
- ❑ Discovery of Pluto

6. Einstein
- ❑ Special Relativity
- ❑ Mass and Energy
- ❑ Space and Time
- ❑ Twin Paradox
- ❑ Absolute Speed Limit
- ❑ General Relativity

Keywords / Precursors to Astronomy

solstice
Long Count
Almagest
Chinese "guest star"
supernova
Crab Nebula
eclipses
heliocentric
Solar System

Keywords / The Copernican Revolution

heliocentric
geocentric
retrograde motion
direct motion
epicycle
deferent
Copernican Revolution
stellar parallax
supernova
comet
ellipse
foci
eccentricity
major axis
semimajor axis
minor axis
Kepler's Three Laws of Planetary Motion
perihelion
aphelion
astronomical unit

Keywords / The Modern Synthesis

telescope
sunspots
Galilean moons
phases
Milky Way
inertia
dynamics
friction
vector
velocity
acceleration
force
Newton's Three Laws of Motion
universal law of gravitation
gravity
universal gravitational constant
center of mass
binary stars
mass
weight
gravitational acceleration
conic sections
parabola
hyperbola
perturbations
special theory of relativity
general theory of relativity

Exercise 2-1: Introductory Narrative

The 1) ____________________ astronomers Aristotle and Ptolemy hold very important places in the history of astronomy. Aristotle proposed that the heavens were composed of 55 2) ____________________ spheres, with the Earth immobile at the center. Planets were attached to the spheres that rotated at constant angular 3) ____________________. To explain retrograde, motion Ptolemy (circa 140 A.D.) proposed that planets were attached to 4) ____________________, which were then attached to the concentric spheres. Although these ideas were incorrect, they became incorporated with Christian theology and were widely believed for almost 2000 years.

In the 16th century, Copernicus proposed a model known as the 5) ____________________ system, where the sun was the center of the Solar System. This system easily explained the varying brightness of planets and retrograde motion using geometrical arguments. However, Copernicus incorrectly retained uniform circular motion in his model.

The correct description of planetary motion came about due to the work of Brahe and Kepler. Brahe designed extremely precise astronomical measuring instruments and collected extensive data on the planet 6) ______________________. Kepler used this data to construct our present model of the solar system which is described by Kepler's Three Laws of Planetary Motion. Kepler eliminated circular motion in his first law by showing that planetary paths are 7) ______________________. He also eliminated constant angular velocity in his second law by showing that planets move 8) ______________________ when they are closer to the sun. His third law relates the 9) ______________________ of the period of revolution of a planet to the 10) ______________________ of its semi-major axis. Thus, an understanding of the true nature of planetary motion was achieved through the combined efforts of Brahe (the greatest observer of his time) and Kepler (the greatest theoretician of his time).

By the early 1600s the teachings of Aristotle and Ptolemy had been accepted for 1500 years, and there was considerable resistance to their being replaced by the 11) _______________ model. Galileo was a pivotal figure in this process, by using the 12) ___________________ to make observations that strongly supported the new theory. His observations of 13) _______________ contradicted Aristotle's view that the heavens were perfect and unchanging. He observed 14) __________ going through a complete set of phases, while in the 15) _____________ system it should always be near crescent phase. His observations of the four large moons of 16) __________ showed that a satellite could orbit a moving planet and not be left behind.

Isaac Newton built upon the work of Galileo and showed that the heavens could be understood in terms of fundamental physical laws. His Three Laws of 17) __________ explain the interplay between force and acceleration. His Universal Law of 18) ______________ states that the force between two objects is proportional to their masses and inversely proportional to the square of their separation. Newton reasoned that this force is what allows the moon to orbit the Earth. He also showed that his laws could be used to derive 19) ____________ Laws of Planetary Motion.

The study of gravitational effects continues today in modern astronomy. The planets Neptune and Pluto were discovered by gravitational 20) _____________ of their orbits by other nearby planets. Einstein's theory of 21) ___________ relativity details how gravity can actually bend the path that light takes. This can be seen in the apparent change of position of stars very near the sun and in the 22) ______________ lensing of distant galaxies.

Exercise 2-2: Exploring Archaeoastronomy*

In this exercise we will think about many of the motions of objects in the sky that greatly interested ancient peoples. Many civilizations left markers and records behind documenting their interest in these motions. It is difficult to generalize when discussing these markers, since every civilization was most concerned with a different aspect of the sky, and many different mechanisms were used to record data concerning it. Thus, the following exercise represents a crude generalization of a primitive astronomical observatory where contributions from many different cultures have been combined. The goal of this exercise is to get you to think about these issues. There really aren't any right or wrong answers.

Imagine that while hiking in a remote area of the Rocky Mountains you come upon an interesting pattern of large stones. The stones are very heavy, and you notice that there are crude paintings on some of the stones. The paintings are weathered, making you reason that the structure is fairly old. You climb a tree to get a better perspective and complete a sketch of the pattern of stones with the top of the paper pointing north. You then climb down and add representations of the paintings on some of the stones. Your final sketch is shown on the following page.

What do you think the pattern of stones and paintings might mean? How might they have been useful to an ancient people?

____

____

____

____

____

____

____

____

____

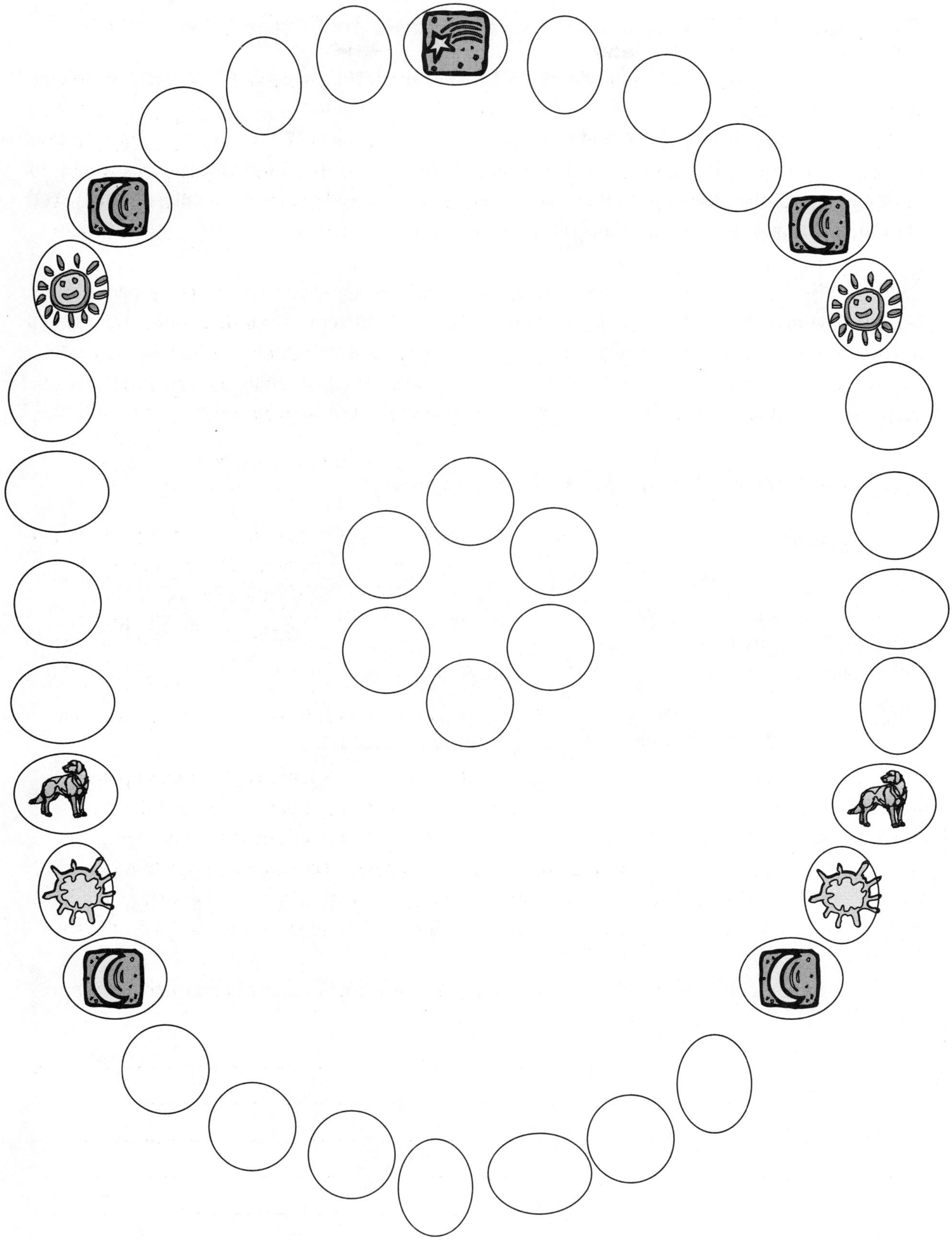

Exercise 2-3: Applications of Kepler's Laws to Other Solar Systems

This exercise will make use of the **Kepler's Laws Calculator (IC 2.36)**, found in the module *Kepler's Third Law*.

This applet takes into account the mass of the orbiting object. We will use this Java Applet to simulate the orbital parameters of more massive objects orbiting stars than the planets of our solar system. In the case of a double star, both stars orbit about a common center of mass. You should note that other solar systems obey Kepler's Laws exactly as our solar system does.

Simulation #1:

Sirius is the brightest star in the sky, and it has a faint companion star that can be seen through small telescopes. The mass of Sirius is 2.2 solar masses, while its smaller companion has a mass of 0.9 solar masses. The separation of the two stars is approximately 20 AU. Use the Kepler's Third Law Calculator to simulate the orbits of the two stars and solve for the orbital period.

Orbital Period of Sirius A and Sirius B = ______________________

Simulation #2:

The binary pair Kruger 60 is only about 13 light years away from our solar system and is one of the best known visual binaries. The components of the system have masses approximately equal to 0.3 (really 0.27) and 0.2 (really 0.16) solar masses. The period of the orbit has been determined visually to be 44.67 years. Use the Applet to determine the separation of the two stars.

Separation of Kruger A and Kruger B = ______________________

Exercise 2-4: Manifestations of Newton's Laws*

Newton's First Law: Every object in a state of uniform motion tends to remain in that state of motion unless an external force is applied to it.
Newton's Second Law: The relationship between an object's mass *m*, its acceleration *a*, and the applied force *F* is $F = ma$. Acceleration and force are vectors; in this law the direction of the force vector is the same as the direction of the acceleration vector.
Newton's Third Law: For every action there is an equal and opposite reaction.
Newton's Law of Gravity: Every object in the Universe attracts every other object with a force directed along the line of centers of the two objects that is proportional to the product of their masses, and inversely proportional to the square of the separation between the two objects.

Newton's Three Laws of Motion and the Universal Law of Gravity are listed above. For each of the factual statements concerning observable phenomena below, identify which of the four laws above most clearly explains why it occurs.

_________________ 1. A force acting on a Ping-Pong ball causes a much larger acceleration than the same force acting on a bowling ball.

_________________ 2. If Jupiter was moved twice as far away from the sun, the force of the sun's gravity on it would be one-fourth as large.

_________________ 3. When the Earth exerts a gravitational force on you, you are exerting an equal and opposite force on the Earth.

_________________ 4. The Voyager 1 space probe has left our solar system (after a gravitational slingshot from Saturn) and will coast at its present speed indefinitely.

_________________ 5. The Earth's gravity produces a constant acceleration of 9.8 m/s^2 downward toward the center of the Earth.

_________________ 6. If there are no forces acting upon an astronaut drifting in space, he or she will move in a straight line, at constant speed forever.

_________________ 7. When firing a gun, you feel a "kick."

_________________ 8. It takes a larger force to give a car an acceleration of 8 m/s^2 than it does to give it an acceleration of 6 m/s^2.

_________________ 9. If we step off a boat onto a dock, the boat tends to move in the opposite direction.

_________________ 10. An air hockey puck set into motion will bounce around the air hockey table for a long time.

Exercise 2-5: The Phases of Venus

In this exercise we will draw the phases of Venus that result from the Ptolemaic and Copernican models of the Solar System. Galileo observed with his primitive telescope that Venus went through nearly a full cycle of phases. Through geometric considerations, we will try to determine whether this is consistent with one model, both, or neither.

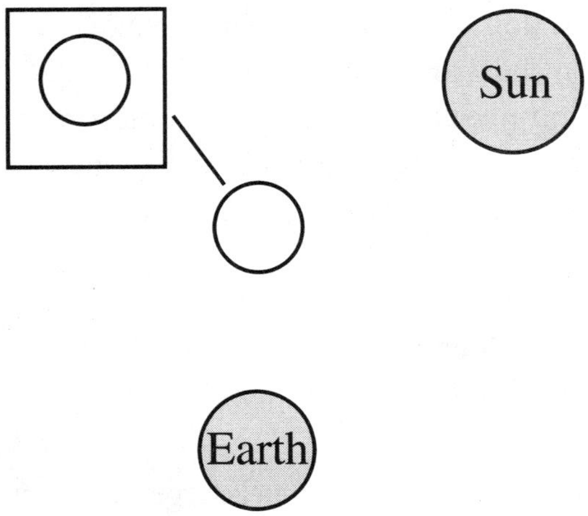

The diagram to the left shows an example configuration for the Earth, sun, and Venus, looking down onto the plane of the Solar System. Note that Venus is shown twice, where the single circle represents the location of Venus in space. We will use the circle in the box as the place where we will sketch in the appearance of Venus as seen from the (North Pole of the) Earth for this configuration.

To determine the phase of Venus from the Earth, first draw in line A. This is simply a line from the center of the sun to the center of Venus. Next, draw in line B. This line is perpendicular to line A and effectively cuts Venus in half. The half of Venus facing the sun is illuminated, while the half of Venus facing away from the sun is in darkness and gets shaded in our diagram.

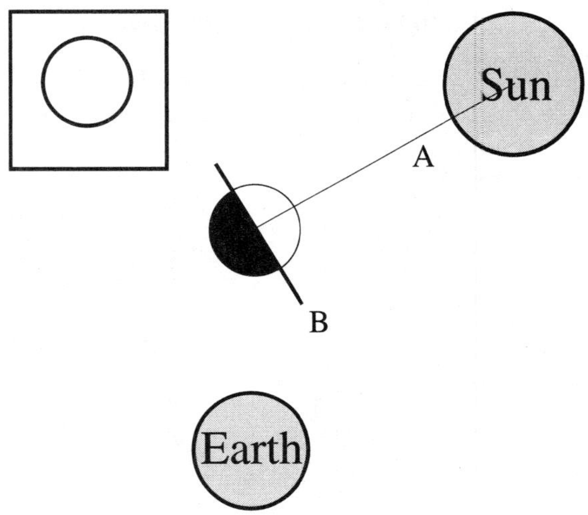

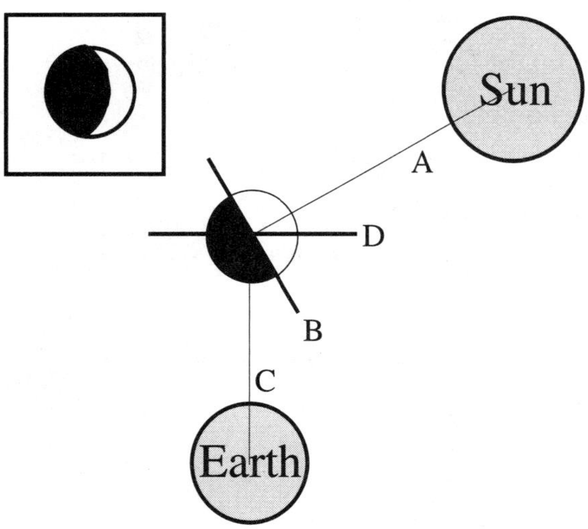

We can now determine the appearance of Venus from the Earth. Draw in line C, which goes from the center of the Earth to the center of Venus. Then draw in line D, which is perpendicular to line C and cuts Venus effectively in half. Line D now defines the appearance from the Earth. For the example to the left, note that most of Venus is dark as seen from the Earth, but there is some illumination on the right-hand side. The line which divides light and dark on the surface is known as the terminator, and should always include the north and south points on the planet as seen from the perspective of the Earth.

Complete six sketches of the appearance of Venus for each of the following two diagrams that illustrate the Ptolemaic and Copernican Geometries. Then, using your sketches, answer: "Which model is consistent with Galileo's observations of a full cycle of phases for Venus?"

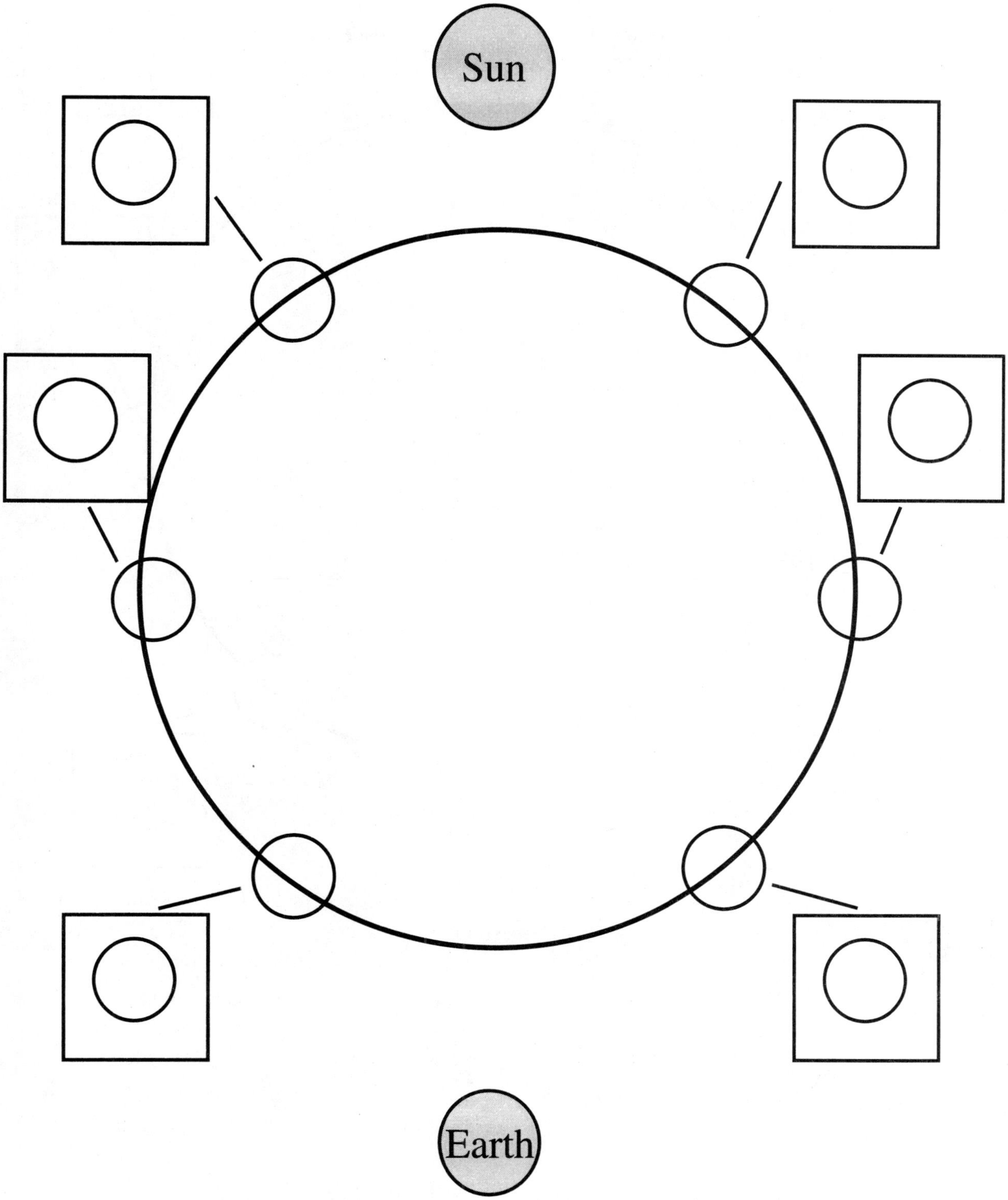

The Phases of Venus in the Ptolemaic System

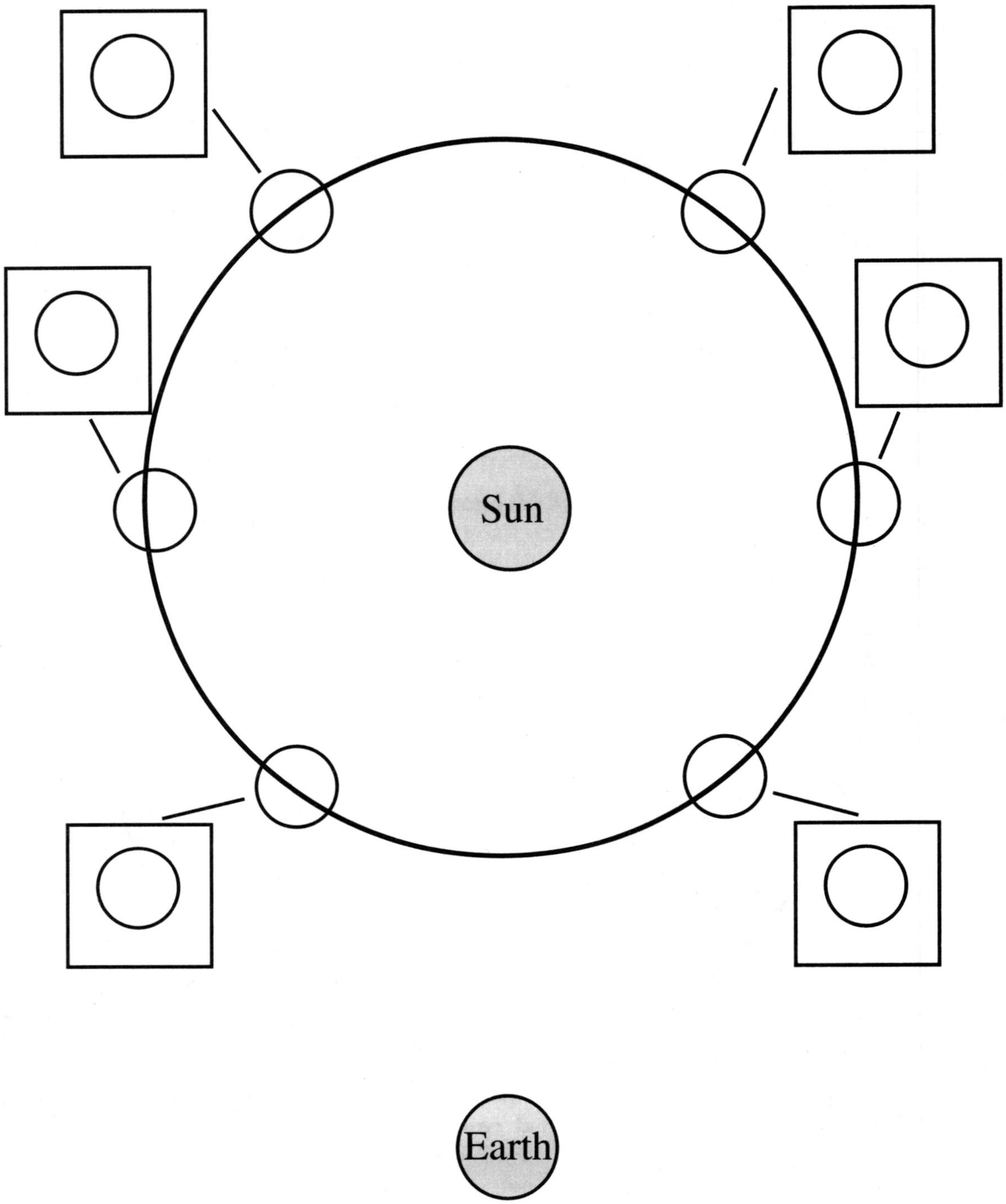

The Phases of Venus in the Copernican System

You can easily check your answers to this exercise using (**IC 2.11**).

Exercise 2-6: True/False Questions

T / F 1. The Big Horn Medicine Wheel in Wyoming indicates the direction of sunrise and sunset on the summer solstice.

T / F 2. The winter constellation Orion plays an important part in the Mayan creation myth.

T / F 3. Aristotle believed that the only motion that heavenly objects were allowed to make was uniform circular motion.

T / F 4. Ptolemy used epicycles to explain retrograde motion.

T / F 5. Aristarchus presented a sun-centered solar system around 2000 B.C.

T / F 6. Eratosthenes measured the circumference of the Earth by sailing one quarter of the way around and multiplying this distance by 4.

T / F 7. Due to the writings of Thomas Aquinas, the teachings of Aristotle and Ptolemy became incorporated into Christian theology.

T / F 8. Copernicus believed in the geocentric theory that the Earth was the center of the solar system.

T / F 9. In the heliocentric theory retrograde motion is explained naturally as due to a changing perspective.

T / F 10. The ancient Greeks could not observe a parallax for stars because the Earth moves so slowly in its orbit.

T / F 11. Tycho Brahe's most important contribution to astronomy was in making theoretical arguments that validated the heliocentric theory.

T / F 12. Since Tycho Brahe found no parallax for either a supernova or a comet he concluded that they were distant objects.

T / F 13. An ellipse is made up of points for which the sum of the distances from two points called foci is a constant.

T / F 14. The eccentricity of a circle is larger than the eccentricity of an ellipse.

T / F 15. The sun is located at the center of a planet's elliptical orbit.

T / F 16. Kepler's second law states that planets move faster when near perihelion.

T / F 17. Kepler's third law suggests that the orbital period of Pluto should be much larger than the orbital period of Mercury.

T / F 18. Galileo was the inventor of the telescope.

T / F 19. From his observations of sunspots Galileo knew that the sun was rotating.

T / F 20. Galileo's observations of the moons of Jupiter proved that the Earth could be moving and our moon wouldn't be left behind.

T / F 21. On the Moon, where there is no atmosphere, a rock would still fall faster than a feather as it does on the Earth.

T / F 22. The nuclei of atoms are held together by the electromagnetic force.

T / F 23. Your mass on the Moon is only about one-sixth of your mass on the Earth.

T / F 24. Kepler's Laws are consistent with Newton's Laws of Motion.

T / F 25. Objects that orbit our sun in parabolic or hyperbolic orbits will only do so once.

T / F 26. The existence of the planet Neptune was predicted by observing its gravitational perturbations on Uranus.

T / F 27. As objects approach the speed of light, they appear to a stationary observer to lengthen in the direction of motion.

T / F 28. Relativity theory predicts that the speed of light is the fastest possible speed.

T / F 29. When gravitational lensing occurs, the actual positions of distant objects are changed due to the influence of gravity.

T / F 30. When one slices through a cone horizontally with a plane, the plane intersects the edge of the cone in the shape of an ellipse.

T / F 31. The stars that are near our Solar System in space exert a strong gravitational force on the planets of our solar system.

T / F 32. Special relativity is only valid for systems that are accelerating.

Conceptual Map 1
The History of Astronomy

In this assignment we will complete a timeline of information detailing the history of astronomy. This is the first of several assignments of this type known as Conceptual Maps, where we will assimilate the information from several chapters into a diagram that will provide a graphical summary. The organizational structure will be provided for you and you will be asked to add information to complete the map.

The two-page format of the Conceptual Maps is especially suited to describing the history of astronomy. The Greeks and other ancient astronomers made many contributions to our knowledge of astronomy. After the year 200 a.d., however, very few contributions were made and considerable information was lost until the time of the renaissance. Thereafter, important contributions were regularly made, accelerating up until the present day.

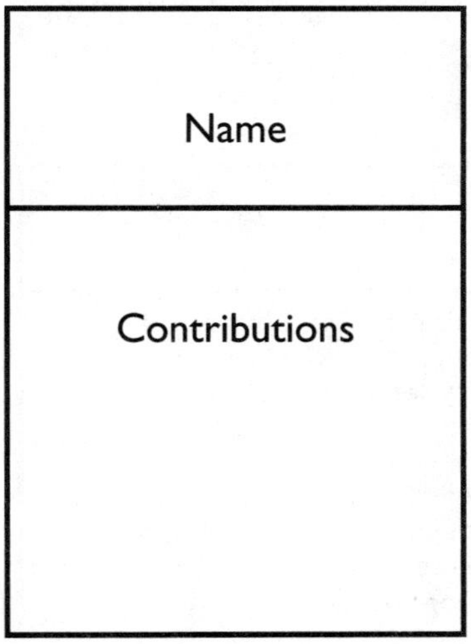

The Conceptual Map is a series of boxes, each with a small area on top to contain the name of an influential astronomer, and a larger area below to list the important contributions of that astronomer. In each of these boxes either the name or the contributions are already completed for you and you are expected to add the other component. A line is drawn connecting the box to the appropriate date on the timeline, when the major contributions of that astronomer were made.

The last page of this section contains some suggestions for adding more information to the map. This will allow you to link the developments in astronomy to the major political, historical, and cultural events going on in the world. These suggestions are very flexible, allowing you to personalize your Conceptual Map and add the information that you find the most interesting and are most likely to remember.

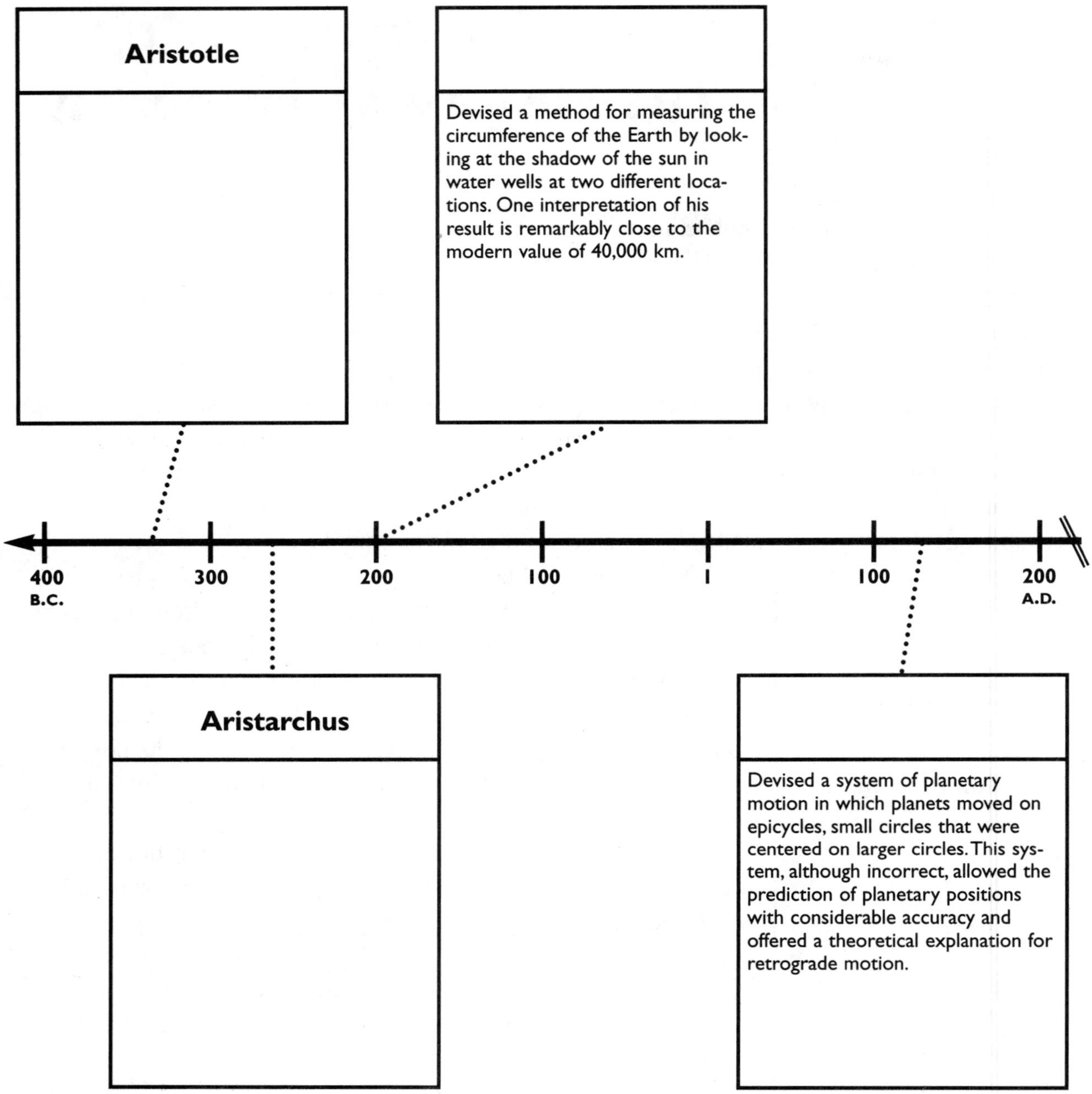

Developing Your Conceptual Map

Here are some suggestions for adding more information to your Conceptual Map.

Add the names of other important astronomers to the timeline. Use the internet and the links provided in *Online Journey Through Astronomy* to locate information on their contributions and the corresponding timeline dates.

- Several figures you could add from ancient times include Thales of Miletus, Hipparchus, Euclid, and Pythagoras.
- For later times, you could add Tycho Brahe, Giordano Bruno, and William Herschel.

Add several important figures and events from history to your timeline. This will help you to place the major discoveries in astronomy within the context of history.

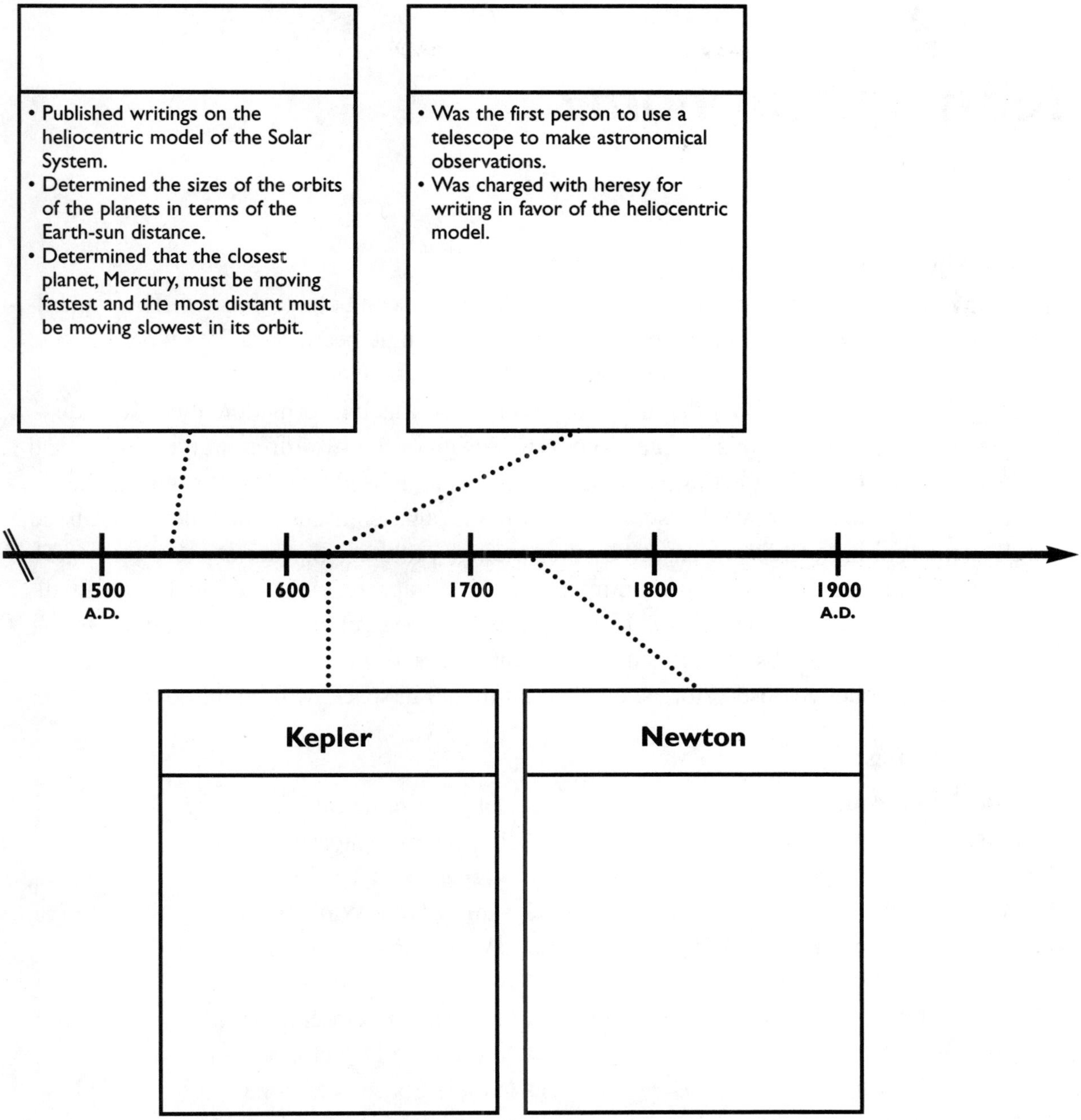

- From ancient times you could add Alexander the Great, Jesus Christ, Julius Caesar, or any other figures you consider important.
- From later times you could add Christopher Columbus' discovery of America, the signing of the Declaration of Independence, the defeat of the Spanish Armada, the landing of the Mayflower, or other events you consider important.

Add several important cultural figures to your timeline.

- You could add the writers Homer and Shakespeare.
- You could add the musician Johann Sebastian Bach.
- You could add the artists Leonardo da Vinci and Rembrandt van Rijn.

Unit 3
Tools of Astronomy

Chapter Objectives

Astronomy is a science in which the objects studied can be seen but not touched. The capability to squeeze every piece of information possible from a single beam of radiation coming to us from a distant star or galaxy is vital if our knowledge of the Universe is to grow and improve. The tools of astronomy that allow us to amass so much information about such distant objects are described in this chapter. The basic design of the two different types of optical telescopes (refracting and reflecting) and the advantages and disadvantages of each will be illustrated. The atmospheric windows through our planet's gaseous envelope will be discussed, as well as the limiting conditions caused by the "seeing" problems this atmosphere causes. The history and development of radio astronomy and its most important spectral line (21 cm) will be described. The new windows on the Universe that have been opened with our ability to travel in space and place observatories in Earth orbit will be discussed. Some of the most important and productive observatories, both on Earth and in space, will be introduced.

Progress Checklist

1. Optical Telescopes
- ❑ Limitations of the Human Eye
- ❑ Refracting Telescopes
- ❑ Reflecting Telescopes
- ❑ Seeing Conditions
- ❑ Adaptive and Active Optic
- ❑ Observatories

2. Instrumentation
- ❑ Photographic Plates
- ❑ Charge-Coupled Devices
- ❑ Spectrographs
- ❑ Computers in Astronomy

3. Radio Telescopes
- ❑ Radio Frequency Observations
- ❑ Neutral Hydrogen
- ❑ Long-Baseline Interferometry
- ❑ Radio Observatories

4. Spaced Based Telescopes
- ❑ Atmospheric Windows
- ❑ UV and IR Observations
- ❑ X-Ray Astronomy
- ❑ Gamma-Ray Observatories
- ❑ Microwave Detectors
- ❑ Hubble Space Telescope

5. Cosmic Rays and Neutrinos
- ❑ Neutrino Astronomy
- ❑ Neutrino Detectors
- ❑ Cosmic Ray Physics
- ❑ Cosmic Ray Detectors

Keywords

telescope
visible light
response of the human eye
refraction
convex lens
concave lens
focus
reflection
focal point
focal length
Cassegrain focus
prime focus
Newtonian focus
eyepiece
chromatic aberration

seeing conditions
twinkling
resolution
light gathering power
aperture
adaptive optics
Keck telescopes
Subaru telescope
Very Large Telescope (VLT)
charge-coupled devices (CCDs)
pixels
spectrographs
diffraction grating
radio astronomy
radio telescope
neutral hydrogen
spin-flip transition
long baseline interferometry
Very Large Array (VLA)
Very Large Baseline Array (VLBA)
Arecibo Observatory
National Radio Astronomy Observatory
Max Planck Institute for Radio Astronomy
atmospheric windows
IRAS (Infrared Astronomy Satellite)
COBE (Cosmic Background Explorer)
ISO (Infrared Space Observatory)
Hubble Space Telescope
IUE (International Ultraviolet Explorer)
EUVE (Extreme U.V. Explorer)
radar imaging
(GRO) Compton Gamma Ray Observatory
Beppo-SAX
cosmic rays
neutrinos
Cerenkov radiation

Exercise 3-1: Introductory Narrative

Our exploration of astronomy began with the many interesting observations that can be made with the human eye. However, the naked eye has some 1) ________________ that we would like to overcome using instrumentation.

The 2) ____________________of the eye is limited due to its small size. Thus, astronomers typically use 3) ______________, with either large lenses or mirrors to collect light.

Another limitation is that the eye distinguishes a new image several times a second and cannot store these images for future reference. Photographic plates overcome both of these shortcomings, but suffer from low 4) _______________, since only about 1% of the incident photons are recorded in the image. They also store their data in 5) __________ format, while most computers require digital data. Modern observations are normally made with "electronic cameras" known as 6) __CCD______. These have efficiencies as high as 7) ___________ and store data in a digital format, allowing it to be used in the future. They are commonly used outside of astronomy today in devices such as camcorders.

The eye also has a limited 8) __________ response, being sensitive only to visible light. There is also a(n) 9) ____________________ in the radio part of the spectrum, in which observations are made from the surface of the Earth with radio telescopes. Wavelengths other than visual and radio must be observed with 10) __________________ telescopes such as the IUE and ROSAT.

When we look at stars from the Earth they appear to be 11) ___________ due to air currents in the atmosphere. Thus, although it isn't necessary, it is often preferable to make visual observations above the atmosphere. The Hubble Space Telescope has extremely good 12) ____________ since its images are not distorted by the turbulence in the atmosphere.

Exercise 3-2: Narrative—Telescope Vocabulary*

The two major types of telescopes are refractors and 1) _______________. The major difference is that refractors are made from lenses while reflectors are made out of 2)_________________. The largest optical telescopes are 3)_____________, since it is easier to build and support large mirrors of high quality.

Refractors are based on the principle of 4) _____________, the bending of light at the boundary between glass and air. Because different wavelengths of light bend by different amounts, when stars are viewed through refractors they are surrounded by fuzzy colored halos. This effect is called 5)_________________________. The lenses used in refractors are called 6)_______________, meaning they are thicker in the middle than near the edges.

Reflecting telescopes work by the principle of reflection, meaning that the angle of 7)______________ of the light is equal to its angle of reflection. The telescope image may focus at a number of different locations known as focus arrangements. Light leaves the telescope through a hole in the primary mirror in a 8)_______________ focus. The observer can actually sit inside the telescope in a 9)_____________ focus. A third focal arrangement, called 10)________________, has light leaving the telescope through a hole in the side of the tube at a 90° angle to the direction from which it came.

There are several criteria by which the ability of a telescope is evaluated. One of these is called the light gathering power. This is simply the area of the mirror, which increases as the 11)________________ of the diameter of the mirror. Another important factor is a telescope's 12)_________________, its ability to separate two nearby objects in the sky. A telescope's ability to make an image larger, which is known as magnification, is not a particularly important capability.

Exercise 3-3: Powers of a Telescope*

Astronomers normally evaluate a telescope based on three properties or powers: light gathering power, resolution, and magnification.

Light Gathering Power is related to the surface area of the objective of a telescope. Catching light in a telescope is just like catching rain in a bucket, the bigger the bucket the more rain. We can compare the light gathering power of two telescopes by taking the ratio of their area.

$$\frac{(LGP)_A}{(LGP)_B} = \left(\frac{D_A}{D_B}\right)^2$$

Thus, if we wanted to compare the light gathering power of a 25 cm cassegrain reflector to the light gathering power of your eye (typical pupil diameter of 7mm), we would get:

$$\frac{(LGP)_A}{(LGP)_B} = \left(\frac{D_A}{D_B}\right)^2 = \left(\frac{25cm}{0.7cm}\right)^2 = 1275$$

Thus, this telescope gathers 1275 times as much light as a typical eye.

Astronomers are also very interested in resolving power, this is the ability to see small detail and is usually given in arc seconds. The resolving power can be roughly calculated using Dawe's Limit:

$$\alpha = \frac{11.6}{D}$$

Where the diameter, D, is given in centimeters and the resolution is in arc-seconds. We can calculate the resolving power of the 25 cm telescope mentioned above as

$$\alpha = \frac{11.6}{D} = \frac{11.6}{25} = 0.46 arc\text{-}sec\text{ onds}$$

Although Dawe's Limit indicates that larger telescope should be able to resolve angles, this isn't true for telescopes on the Earth. The limiting factor is not the telescope but the blurring due to the air currents in the atmosphere. The best possible resolving power from the ground is often cited as 0.5 arc-seconds.

Telescope manufacturers often cite the magnification of a telescope as well. However, this isn't a particularly important quantity. It is far more important to collect light for faint objects and resolve detail than to make the image bigger. Magnification is equal to the focal length of the telescope, objective divided by the focal length of the eyepiece.

$$M = \frac{F_O}{F_E}$$

Thus, a particular telescope is capable of a range of magnifications depending on the eyepieces that come with it.

Estimate the actual Light Gathering Power (compared to the eye) and Resolving Power for the following telescopes.

Yerkes 1-m refractor
Light Gathering Power
Resolving Power

Hubble Space Telescope – Diameter = 2.4 m
Light Gathering Power
Resolving Power

Exercise 3-4: True/False Questions

T / F 1. The twinkling of stars is caused by currents in the atmosphere.

T / F 2. Photographic plates are an extremely efficient way to record astronomical images.

T / F 3. The 21-cm line is produced by the electron in a hydrogen atom moving down to a lower permitted energy level.

T / F 4. Atmospheric windows appear in the visual and ultraviolet portions of the electromagnetic spectrum.

T / F 5. Most of the infrared light entering our atmosphere is absorbed by excitations of rotations and vibrations for molecules.

T / F 6. X-rays and gamma rays are important for mapping the location of cool interstellar hydrogen.

T / F 7. The Cosmic Background Explorer (COBE) has detected microwave radiation left over from the Big Bang.

T / F 8. It is possible to connect two radio telescopes and have the resulting resolution of a single telescope of a size equal to their separation.

T / F 9. The HST has very good resolution since the light it collects hasn't passed through the atmosphere.

T / F 10. Neutrinos are very difficult to detect because they interact very weakly with matter.

T / F 11. Direct studies of cosmic rays should ideally be done deep underground.

T / F 12. The electronic cameras that are attached to most telescopes today are called charge coupled devices.

T / F 13. CCDs are almost as efficient detectors of radiation as the human eye.

T / F 14. The light gathering power of a telescope is roughly equivalent to the area of the light collecting surface.

T / F 15. Radio telescopes are good for observing cool clouds of hydrogen.

Unit 4
Interaction of Light and Matter

Chapter Objectives

Almost all of the information we gather concerning our Universe comes to us in the form of electromagnetic radiation. So, understanding the nature of this radiation and the data it brings to us is essential to the science of astronomy. This chapter discusses the different portions of the electromagnetic spectrum and the physical laws and processes that are applicable to this radiation. Planck's Law, Wien's Law, and the Stefan-Boltzmann Law and their consequences will be explained. The causes of refraction, diffraction, and dispersion will be illustrated. The Inverse Square Law of Light will be demonstrated, and then applied, to determine the amount of flux we receive from stars. The Doppler Effect and its usefulness in astronomy will be explained. The structure of the atom will be shown to be the underlying reason for the spectral lines that serve as identification of each of the chemical elements. The Bohr Model of the hydrogen atom will be explored in order to understand the excitation and de-excitation processes that give rise to absorption and emission spectra.

Progress Checklist

1. Electromagnetic Radiation

- ❑ Light as a Wave
- ❑ Electromagnetic Spectrum
- ❑ Refraction and Diffraction
- ❑ Dispersion of Light
- ❑ Doppler Effect
- ❑ Inverse Square Law

2. Radiation Laws

- ❑ The Planck Radiation Laws
- ❑ The Stephan-Boltzmann Law
- ❑ The Wien Radiation Law
- ❑ Planck Law Java Applet
- ❑ Stefan-Boltzmann Java Applet
- ❑ Black Body, the Game

3. Atoms & Molecules

- ❑ Components of the Atoms
- ❑ Isotopes
- ❑ Periodic Table
- ❑ The Bohr Atom
- ❑ Ionization and Plasmas
- ❑ Molecules and Molecular Spectra

4. Spectra

- ❑ Spectrographs and Spectra
- ❑ Origin of Spectra
- ❑ The Hydrogen Spectrum
- ❑ The Zeeman Effect

5. Non-Thermal Emission

- ❑ Nonthermal Emission
- ❑ Synchrotron Radiation
- ❑ Polarization
- ❑ Synchrotron Power Supply

Keywords

electromagnetic wave
electromagnetic spectrum
Ångstrom
photon
frequency
wavelength
Planck's constant
gamma rays
X-rays
ultraviolet
visible light
infrared

microwave
radio waves
refraction
diffraction
dispersion
Doppler effect
spectrograph
red shift
blue shift
inverse square law
luminosity
flux
blackbody radiation
Planck radiation law
Wien displacement law
Stefan-Boltzmann law
atom
molecule
electron
proton
neutron
nucleus
atomic number
atomic mass
isotopes
periodic table
ion
plasma
Bohr model of the atom
Balmer series
excitation
quantum mechanics

Exercise 4-1: Introductory Narrative

Most of the information astronomers have collected about the Universe has come to Earth in the form of 1) ___________. It comes in different wavelengths or 2) ___________, which collectively are known as the electromagnetic spectrum. Long wavelength photons have relatively little energy and are used to observe 3) ________ temperature objects, while short wavelength photons come from high temperature objects.

When a 4) ________________ is used to separate the light according to wavelength the resulting intensity versus wavelength, graph is known as a spectrum. If all of the wavelengths of light are present it is known as a(n) 5) _____________ spectrum, which is produced by a material whose atoms are close together. This is the type of spectrum produced by a light bulb. This type of spectrum is described by the 6) ___________________ Law, which gives the intensity of radiation emitted per unit surface as a function of wavelength for a given temperature. The wavelength at which the most light energy is radiated is described by 7) ___________ Law and the total amount of light energy radiated per second per square meter is described by 8) _______________________ Law.

Low density substances typically produce a spectra where not all of the wavelengths of light are present. A thin, hot gas will produce a(n) 9) ____________ spectrum (where only selected wavelengths are present), which are produced when excited electrons move back down to the ground state. A thin cool gas produces a(n) 10) _____________ spectrum when a(n) 11) _____________ spectrum is incident upon it. This is the type of spectrum produced by a star. These two types of spectra are produced because an electron can only have certain 12) ____________ values of energy. Thus, only wavelengths of light whose energy corresponds to the energy difference between two energy 13) _____________ are involved.

Exercise 4-2: Energy, Wavelength, and Frequency of E-M Waves*

For each of the following trios of photons from the electromagnetic spectrum, complete the table by identifying whether photon A, photon B, or photon C has the largest wavelength(λ), largest frequency(f), and the largest energy(E).

You will find the following equations involving frequency(f), wavelength(λ), and the speed of light(c) useful.

$$E = \frac{hc}{\lambda} = h\nu \qquad \nu = \frac{c}{\lambda}$$

	Photon A	Photon B	Photon C	Largest Wavelength	Largest Frequency	Largest Energy
1	Violet	Green	Red	C	A	A
2	Yellow	Blue	Orange			
3	Green	Blue	Violet			
4	Visual	UV	Infrared			
5	Microwave	X-ray	Visual			
6	UV	Radio	Infrared			
7	X-ray	Gamma Ray	Visual			
8	Gamma Ray	UV	Radio			
9	Infrared	Radio	Microwave			
10	Visual	X-ray	Gamma Ray			

Exercise 4-3: Energy Level Diagrams*

Here is a graphical representation of the electron energy levels of a hydrogen atom. The first six levels are shown, as well as a free state, indicating that the electron is not bound to the nucleus. Indicate the letter(s) of the transition(s) that are appropriate for each of the following statements.

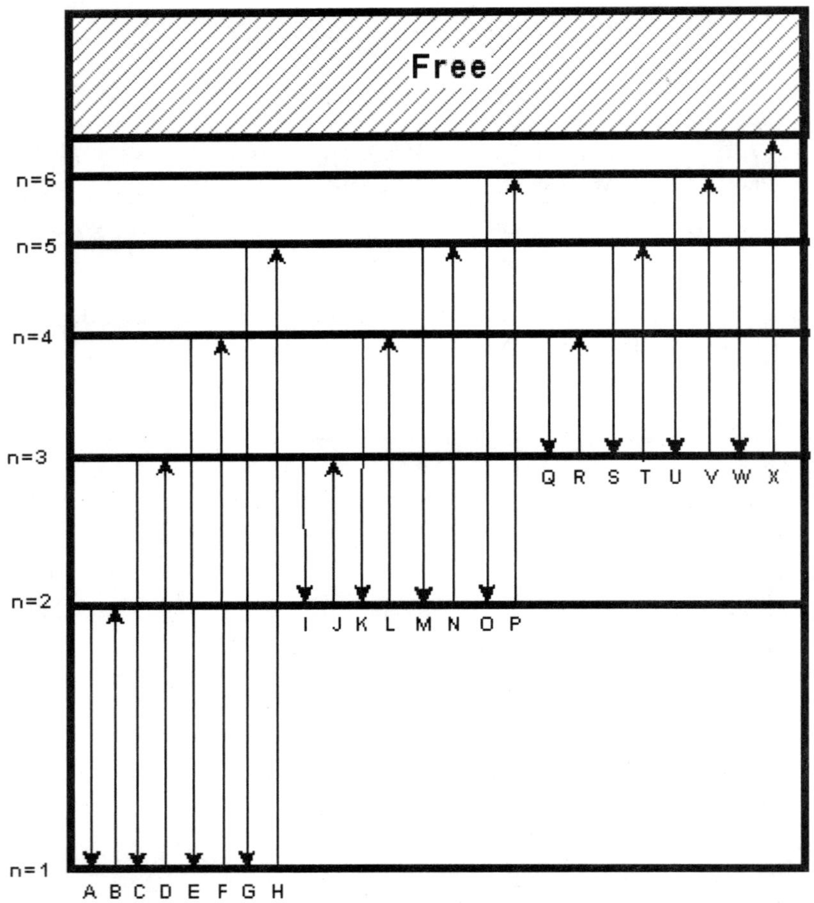

1. IKMO emit(s) visible light
2. _______ absorb(s) infrared light
3. _______ emit(s) UV light
4. _______ emit(s) red light
5. _______ absorb(s) violet light
6. _______ emit(s) Paschen lines
7. _______ emit(s) Balmer lines
8. _______ emit(s) Lyman lines
9. _______ emit(s) Hb
10. _______ absorb(s) Ha
11. _______ emit(s) the longest wavelength
12. _______ emit(s) largest frequency
13. _______ absorb(s) shortest wavelength
14. _______ absorb(s) smallest frequency
15. _______ represents ionization

Exercise 4-4: Applying Radiation Laws to Stars*

In this exercise we will apply Wien's Law and the Stefan-Boltzmann Law to a range of temperatures representative of the temperatures astronomers find on the surface of stars.

Recall that Wien's Law tells us the peak wavelength of the black body spectrum. This is the wavelength at which the most light is being produced. It is expressed mathematically as

$$\lambda_{peak} = \frac{2.9 \times 10^7}{T}$$

Where λ is in Angstroms and T is in Kelvin.

The Stefan-Boltzmann Law allows one to calculate the energy flux, the total energy being emitted at all wavelengths of a blackbody from each square centimeter of the surface for each second. This is analogous to the total area under the Planck curve. The law is written as $E = \sigma T^4$ where $\sigma = 5.67 \times 10^{-5}$ erg-cm^{-2}- K^{-4}- s^{-1}.

If we apply these laws to the sun, which has a rather average stellar surface temperature of 5800 K, we obtain

$$\lambda_{peak} = \frac{2.9 \times 10^7}{T} = \frac{2.9 \times 10^7}{5800K} = 5000 \text{ Å}$$

$$E = \sigma T^4 = \left(5.67 \times 10^{-5} \frac{\text{ergs}}{s - cm^2 - K^4}\right)(5800K)^4 = 6.42 \times 10^{10} \frac{\text{ergs}}{s - cm^2}$$

If we apply these same laws to Sirius, the brightest star in the sky, which has a surface temperature of 9500 K, we obtain

$$\lambda_{peak} = \frac{2.9 \times 10^7}{T} = \frac{2.9 \times 10^7}{9500K} = 3053 \text{ Å}$$

$$E = \sigma T^4 = \left(5.67 \times 10^{-5} \frac{\text{ergs}}{s - cm^2 - K^4}\right)(9500K)^4 = 4.62 \times 10^{12} \frac{\text{ergs}}{s - cm^2}$$

Thus, the hotter star has a peak wavelength at a shorter wavelength (which corresponds to higher energy light) and a larger energy flux, as would be expected from our knowledge of Planck curves.

We can form a ratio to compare the energy flux from each square centimeter of Sirius to that of the sun.

$$\frac{E_{Sirius}}{E_{Sun}} = \frac{\sigma T^4_{Sirius}}{\sigma T^4_{Sun}} = \frac{T^4_{Sirius}}{T^4_{Sun}} = \frac{(9500K)^4}{(5800K)^4} = 7.2$$

Thus each square centimeter of the surface of Sirius produces 7.2 times as much energy as does the surface of the sun. We will use this type of calculation again later in the course when we discuss the luminosities of stars.

Calculate the peak wavelength, energy flux, and the ratio of the energy flux as compared to the sun's flux for the following stars.

Star #1 Spica. This is one of the hottest of the bright stars in the sky with a surface temperature of approximately 25,000 K.
Peak Wavelength
Energy Flux
Ratio of Spica's Energy Flux to Sun's Energy Flux

Star #2 Betelgeuse This is one of the coolest of the bright stars in the sky with a surface temperature of approximately 3,000 K.
Peak Wavelength
Energy Flux
Ratio of Betelgeuse's Energy Flux to Sun's Energy Flux

Exercise 4-5: True/False Questions

T / F 1. Our eyes are most sensitive to yellow light.

T / F 2. If a star is moving toward our Solar System, its spectral lines will be redshifted.

T / F 3. The Zeeman Effect is the splitting of spectral lines due to the presence of strong electrical fields.

T / F 4. The two types of electromagnetic radiation that can penetrate the atmosphere are visual and microwave.

T / F 5. Of two stars giving off electromagnetic radiation, the hotter star will produce more radiation at all wavelengths.

T / F 6. When astronomers observe a typical star using a telescope and spectrograph, they will detect a continuous spectrum.

T / F 7. An atom of singly ionized helium has one electron.

T / F 8. If the distance to a star were doubled, its intensity would decrease by a factor of 2.

T / F 9. A blackbody spectrum increases in intensity at very long wavelengths.

T / F 10. A chart of the elements arranged to show periodic properties is called a periodic table.

T / F 11. The atmosphere is very opaque in the UV because its molecules absorb UV light very strongly.

T / F 12. Strong sources of non-thermal radiation are observed to be associated with active galaxies.

T / F 13. Due to the Earth's velocity of revolution about the sun, we see spectral lines from the sun blueshifted.

T / F 14. A blackbody spectrum produced by an object at 2000°K has a peak wavelength equal to twice the value of a blackbody spectrum at 4000°K.

T / F 15. Atoms can reach an excited state through a collision.

Conceptual Map 2
The Electromagnetic Spectrum

In this assignment we will complete a diagram of information detailing the various wavelength bands of the electromagnetic spectrum and how they are used in astronomy. This is the second of several assignments of this type, known as Conceptual Maps, where we will assimilate the information from several chapters into a diagram that will provide a graphical summary. The organizational structure will be provided for you and you will be asked to add information to complete the map.

The electromagnetic spectrum is divided into seven wavelength regions. The wavelength axis of the diagram is in units of meters and contains seven rectangular boxes which will contain the names of each wavelength region. The X-ray region is already labeled for you.

Each of the seven boxes containing the name of a wavelength region is linked to a larger box that will contain a description of how this wavelength region is used in astronomy. The larger box has a small area on top to contain the name of telescopes or space probes that collect that type of observation, and a larger area below to list the types of astronomical objects that can be studied in that wavelength region. Try to start with a generalization like "very hot gases" before you list specific objects. This box has been completed for you in the X-ray region.

You should return to this Conceptual Map frequently throughout the course. When you learn about a new type of astronomical object in later chapters, try to place it within the context of this diagram.

Uhuru, Einstein, ROSAT, EXOSAT
Very hot gases, violent processes • neutron stars • accretion of binary stars • black holes

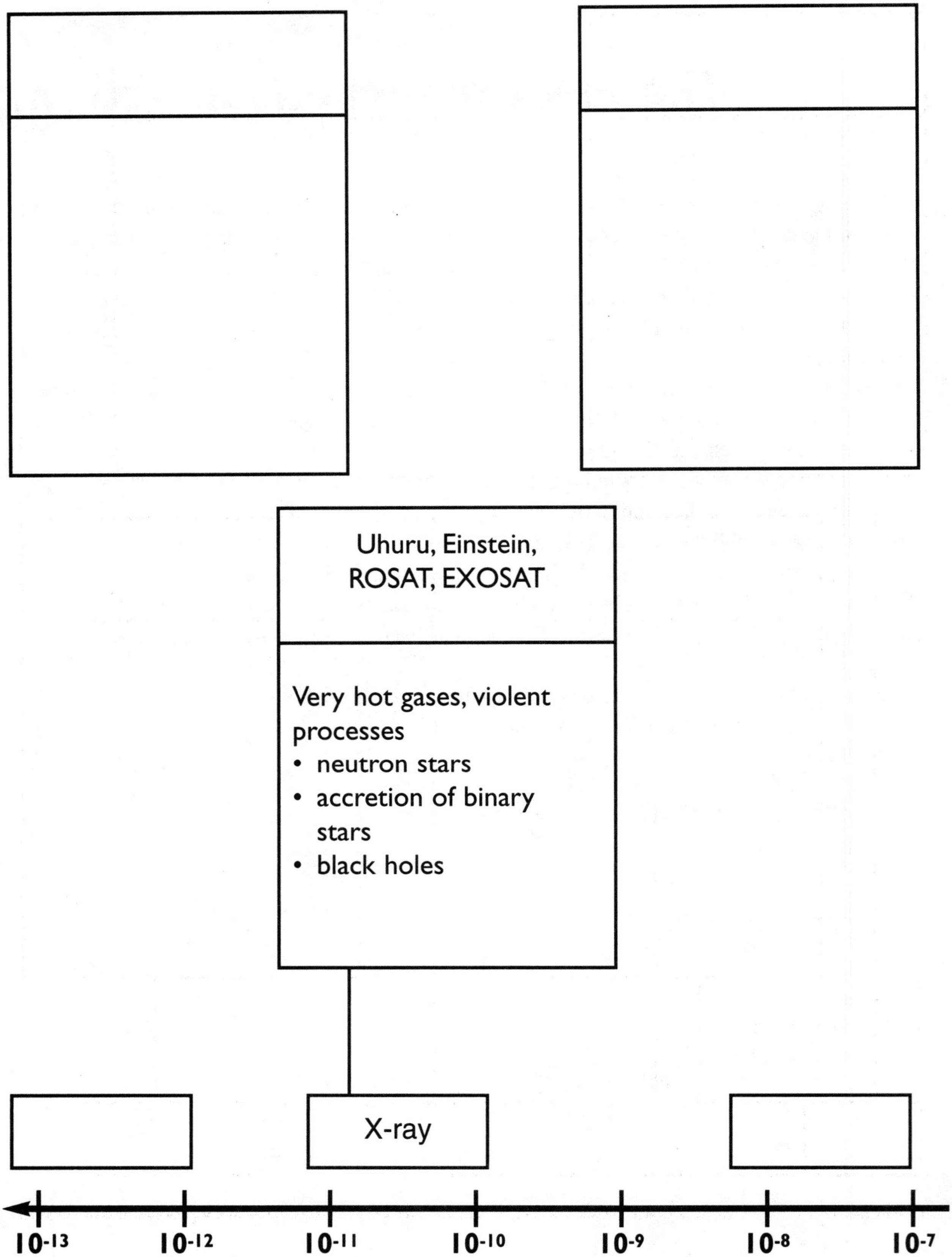
Uhuru, Einstein,
ROSAT, EXOSAT
Very hot gases, violent processes
• neutron stars
• accretion of binary stars
• black holes
X-ray
10-13
10-12
10-11
10-10
10-9
10-8
10-7

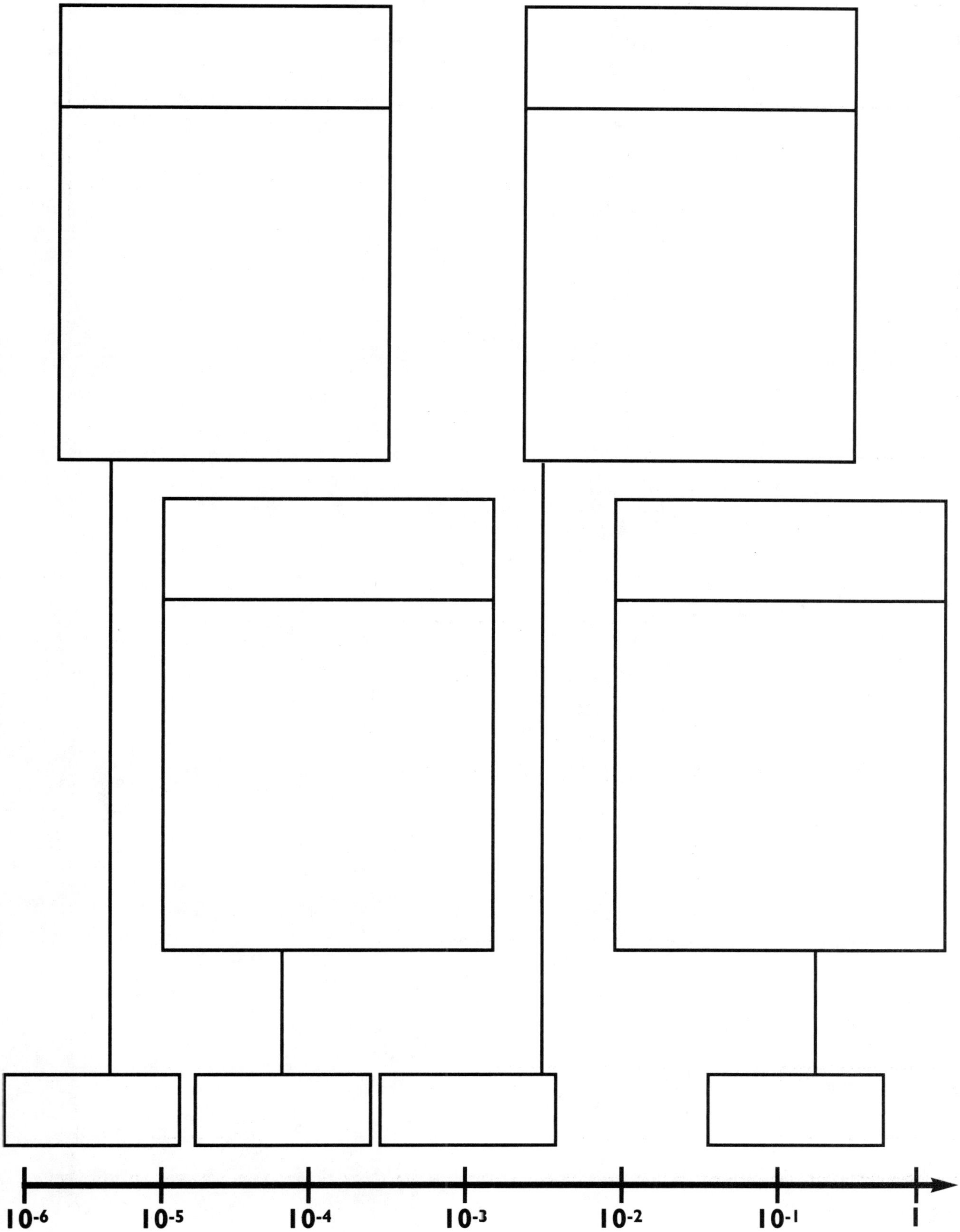
10^{-6}
10^{-5}
10^{-4}
10^{-3}
10^{-2}
10^{-1}
1

Unit 5
Origin of the Solar System

Chapter Objectives

The large scale structure of our Solar System shows regularities in the planets' motions and spacing, and also the division of the planets into two distinct types—the Terrestrials and the Jovians (or Gas Giants). In this chapter we will discuss the sizes, masses, and densities of the Solar System members and how these parameters may have originated. We will apply Kepler's Three Laws of planetary motion and Newton's laws of motion to the planetary orbits in order to understand their regularities. The role of angular momentum in Solar System formation will be discussed.

Progress Checklist

1. Overview of the Solar System

- ❑ The Relative Radii
- ❑ Masses and Densities
- ❑ Orbital Radii
- ❑ Eccentricity and Orientation

2. Formation of the Solar System

- ❑ Angular Momentum
- ❑ Nebular Hypothesis
- ❑ New Star Formation
- ❑ New Solar Systems
- ❑ Planets around other Stars
- ❑ Binary Systems vs. Planetary

Keywords

solar system
Terrestrial planets
Gas Giants
density
revolution
rotation
plane of the ecliptic
Nebular hypothesis
angular momentum
gas pressure
protosun
protoplanet
proplyd
binary stars

Exercise 5-1: Introductory Narrative

There are 1) __________ planets in our Solar System. The orbits of these planets are very nearly in the same 2) ______________, which makes our Solar System shaped like a disk. All of the planets 3) ____________ in the same direction. Pluto and Mercury have fairly large values of 4) ____________, but the orbits of the other planets are very nearly 5) __________.

The characteristics of the planets are quite diverse. The gas giant (or Jovian) planets are very 6) __________ , very massive, and are found in the outer parts of our Solar System. However, they have very low values of 7) ___________, since they are composed primarily of hydrogen and helium. The inner (or terrestrial) planets, on the other hand, are very small and have low masses. They also have solid surfaces.

The 8) __________________ proposes that our Solar System was formed from a collapsing cloud of gas and dust. As the cloud collapsed, its rate of rotation 9) ______________ and it flattened into a disk. The planets then formed from instabilities in this rotating disk of material while the star formed at the center.

Astronomers believe this model is correct for several reasons. They can actually see the flattened disks of material around young stars such as Beta Pictoris. 10) __________ in these disks suggest that planets are sweeping out paths through the dust. Images from the Orion Nebula show the protoplanetary disks at any even earlier stage of development. Astronomers can also indirectly detect planets around ordinary stars using techniques based upon 11) ____________________, the movement of the star around the Solar System's center of mass. The discovery of planets around other stars supports the nebular hypothesis, since in it planet formation is a natural consequence of star formation.

Exercise 5-2: Scale Drawings of Planetary Orbits

Scaled drawings of elliptical planetary orbits can easily be drawn using commonly found objects: a loop of string, a pencil, and two pins or thumbtacks to represent the foci. A piece of cardboard placed behind the paper will allow the tack to hold the paper in a more stable manner. These simple materials and some orbital parameters are all you need to draw accurate scale models of planetary orbits.

To draw an ellipse on the paper, you first need to know where to put the pins and how long to make the string loop. These distances depend on the particular orbit you are trying to draw and the scale factor you choose. The term eccentricity describes the shape of the ellipse. It is defined as the distance between the foci (F_1 and F_2) divided by twice the semi-major axis ($2a$),

$$E = \frac{\overline{F_1F_2}}{2a}$$

You can see that if we move the two foci close together (F_1F_2 is small) the eccentricity gets small. In fact, if the two foci merge together, we have a circle where $e = 0$. On the other hand, if the two foci are very far apart, we get a long, thin ellipse. If they are infinitely far apart we have the mathematical limit of a parabola. Solving for F_1F_2 yields

$$F_1F_2 = 2ae$$

This quantity determines how far apart to set the pins or tacks on your cardboard.

The next step is to pick a planet and decide on a scale to use. In this example we will focus on the orbit of Mars, as Kepler did. The orbit of Mars has a semi-major axis equal to 1.52 AU and an orbital eccentricity, 0.093. Using the mathematical expression given above, you can calculate that the distance between the foci F_1F_2 is 0.28 AU. A convenient scale for this drawing might be 1 AU equals 5 cm. Then the separation between your tacks should be 1.4 cm.

Now you need to prepare your loop of string. (There are two possible approaches that could be used here. You may either tack down the ends of string or have one long loop of string that encircles the two thumbtacks. The latter approach will be used in this example.) The sum of the distances from one focus to any point on the ellipse and then to the other focus is $2a$. However, because there is an extra length of string connecting the pins, the total length of your loop should be $2a + F_1F_2$ or 3.323 AU, which corresponds to about 16.6 cm. You can approximate this length by holding the ends of the string together, putting the loop on a ruler with the middle of the loop at zero, pulling the string taut, and tying a knot at 8.3 cm, half the total length of the loop.

Finally, hook the string over the pins or thumbtacks and, keeping the string taut with the pencil, trace out the ellipse. You now have a scale drawing of the orbit of Mars, which is shown on the next page, for the scale factor of 1 AU equals 5 cm.

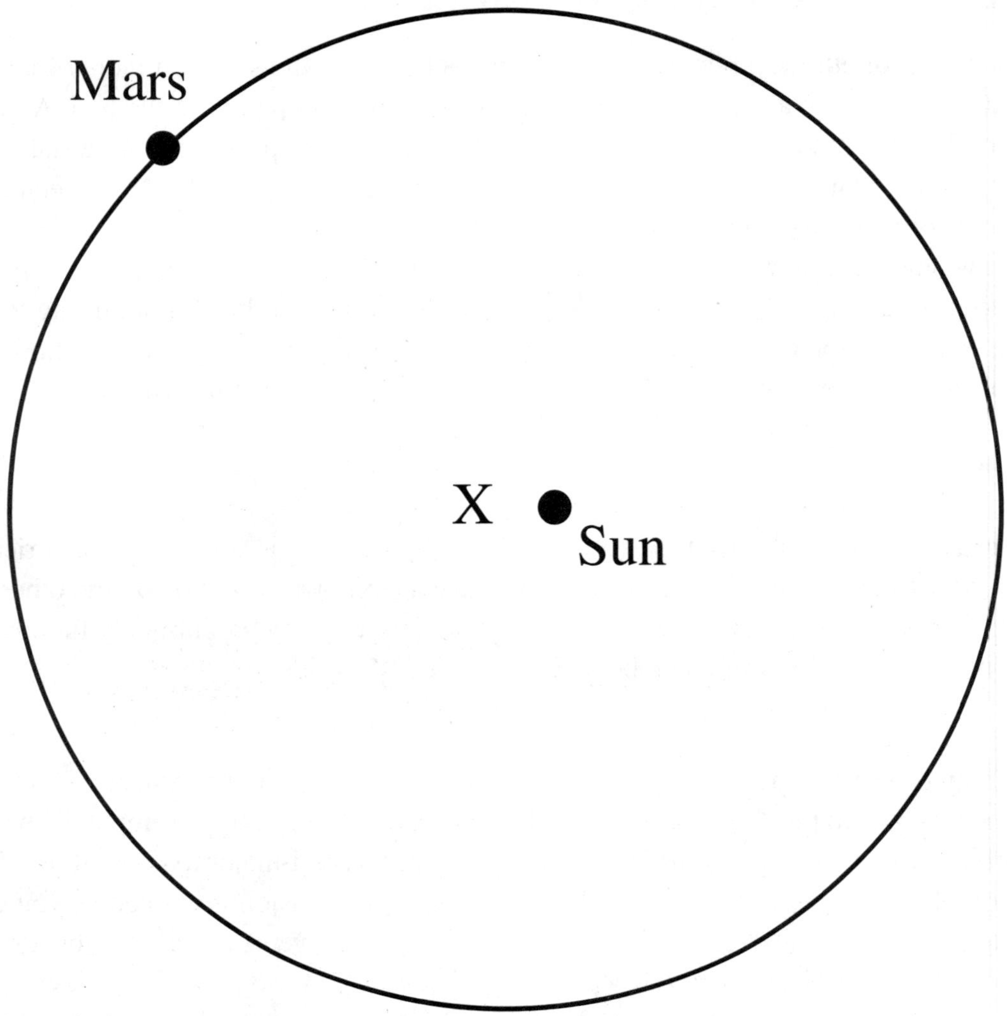

Semi-major axis = 1.52AU
Eccentricity = 0.093
Scale Factor - 1 AU = 0.5 cm
Separation of foci = 0.28AU = 1.4cm
String Length = 3.323 AU = 16.61cm

Your Assignment

Use the technique described on the previous page to create a scale drawing of the orbit of the planet Pluto (a = 39.5 AU, e = 0.248). Clearly label the two foci and draw in the semi-major axis on your diagram. Draw in Pluto and the sun. You should also create a legend detailing the value of the semi-major axis in AU, the eccentricity, the scale factor of your drawing, the value of F_1F_2, and the length of your string. Your scale factor should be chosen so that the drawing fits nicely on a single sheet of paper (but uses most of that single sheet).

Exercise 5-3: Simulating the Orbits of ExtraSolar Planets*

This exercise will make use of the Java applet entitled **Extrasolar Planet Applet (IC 5.5)**. You should have the orbit and grid enabled for this assignment.

Simulation #1: Simulate the orbit of the planet around HD 210277.
Enter the following parameters into the slider bars.

$M_{planet} = 1.3\ M_{Jupiter}$
$M_{star} = 0.9\ M_{\upsilon}$
$a = 1.1$ AU
$e = 0.45$

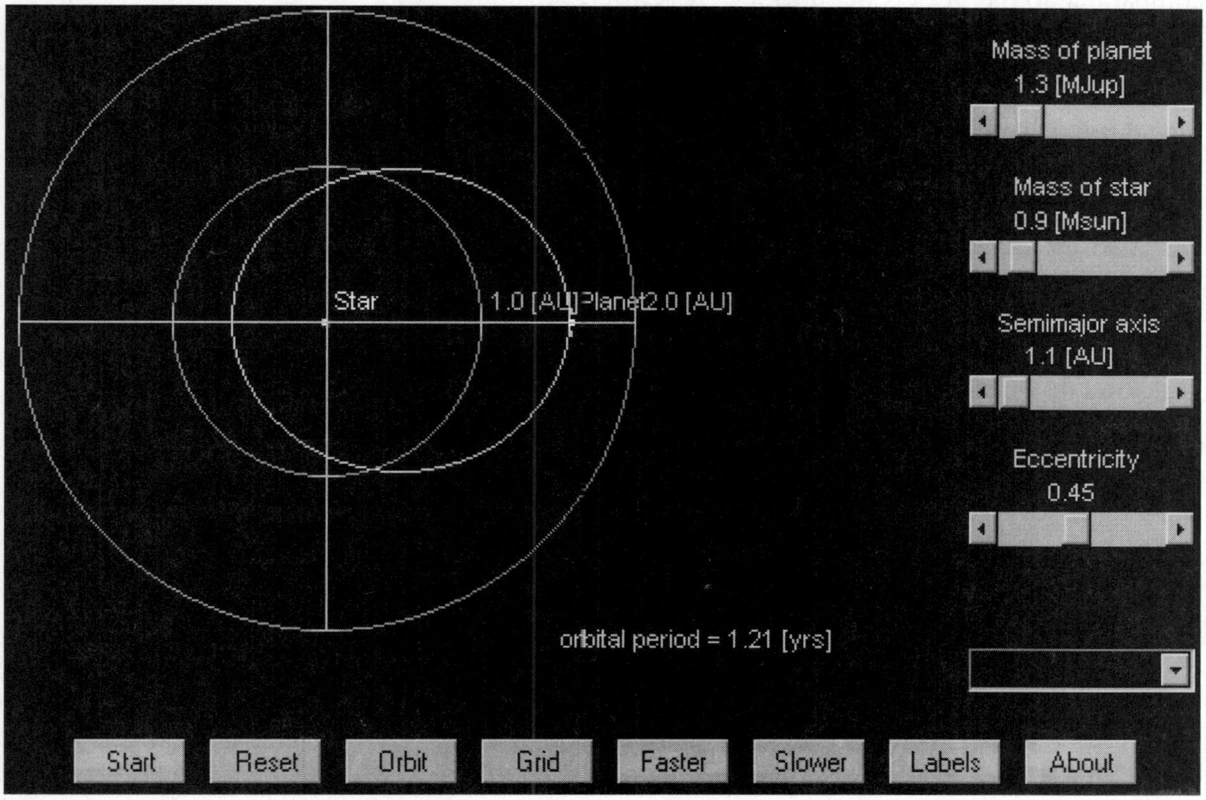

Use the simulation to determine the following for HD 210277:

Period = ________________________

Periastron Distance = ________________________

Apastron Distance = ________________________

Simulation #2: Simulate the orbit of the planet around HD 195019.

$M_{planet} = 3.5\ M_{Jupiter}$

$M_{star} = 1.0\ M_{\upsilon}$

$a = 0.1$ AU

$e = 0.03$

Period = ______________________

Question #1: Would either of these planets fit in well in our Solar System? Why? Why not? (Are the values of planetary mass, eccentricity, and orbital radii for these extrasolar planets comparable to the planets in our Solar System?)

____

____

____

____

Question #2: Can you speculate on a reason for your answer to the above question?

____

____

____

____

Exercise 5-4: True/False Questions

T / F 1. Jupiter orbits the sun with the greatest orbital velocity since it is the most massive planet.

T / F 2. The density of Saturn is greater than the density of Venus.

T / F 3. Saturn is farther from the sun than is Mars.

T / F 4. The orbit of Mercury has an eccentricity of zero.

T / F 5. Pluto has the most eccentric orbit and the largest orbital inclination.

T / F 6. Angular momentum and mass-energy are conserved quantities.

T / F 7. Since Uranus is larger than Neptune but less massive, Neptune must have a greater density.

T / F 8. A collapsing cloud of gas will spin more slowly as it collapses.

T / F 9. A spinning ice skater is an appropriate context within which to discuss the conservation of angular momentum.

T / F 10. The nebular hypothesis predicts that the sun should be spinning much faster than it does today.

T / F 11. There is strong evidence that stars are being born in the Eagle Nebula and the Orion Nebula.

T / F 12. The majority of planets discovered around other stars are just like the planets in our Solar System.

T / F 13. Whether or not planets can orbit binary stars in stable orbits is an unsolved question at this time.

T / F 14. Stars occur most commonly as single stars like the sun.

Unit 6
The Earth and Moon

Chapter Objectives

Knowledge of the basic physical features of the Earth and an understanding of how the Earth has changed since its formation will be essential in our study of the Solar System. In this chapter we will examine the physical makeup of the Earth and its atmosphere. Throughout this course, we will use our home planet as the basis of comparison for understanding the composition and evolution of the other terrestrial planets; and Earth's weather patterns will be compared to the more intense meteorology of the Jovians. In this chapter we will catalogue and attempt to explain Earth's properties, its interior structure, its history, and its atmospheric dynamics. Some of the scientific tools used to obtain this knowledge, such as seismology, will be introduced.

Our Moon serves as a "stepping stone" for our spacecraft exploration of the entire Solar System. This chapter will first describe the information we were able to deduce concerning our satellite before the Apollo missions. The wealth of data the Apollo seismic studies and detailed lunar rock analyses added to our understanding of our nearest neighbor in space will be discussed. The general properties of our Moon in terms of size, mass, density, and overall composition will be compared to those of Earth. The lunar surface features and the history of the Moon that led to the development of these different features will be described. The reason that the crater count on the Maria is so low compared to that of the Highlands will be illustrated. The various hypotheses of how our Moon may have originated will be presented, and the problems and constraints for each hypothesis discussed.

Progress Checklist / The Earth

1. Seismic Waves
- ❑ Seismic Waves
- ❑ Interior of the Earth
- ❑ Inner and Outer Cores
- ❑ Geological Differentiation

2. Plate Tectonics
- ❑ Evidence for Plate Tectonics
- ❑ Crustal Plates
- ❑ Motion of the Crustal Plates
- ❑ The Asthenosphere
- ❑ Convection Currents
- ❑ Consequences of Plate Tectonics

3. Atmosphere
- ❑ Composition
- ❑ Origin
- ❑ Troposphere
- ❑ Stratosphere
- ❑ Mesosphere and Ionosphere
- ❑ Ozone Layer

4. Weather
- ❑ Complex Weather Patterns
- ❑ Consequences of Rotation
- ❑ Solar Heating
- ❑ Cyclones and Anticyclones
- ❑ Weather Fronts
- ❑ Realistic Weather Patterns

5. Oceans
- ❑ Heat Capacity and Latent Heat
- ❑ Ocean Currents
- ❑ Currents and Climate
- ❑ Seawater Chemistry
- ❑ Sea Surface Temperatures
- ❑ El Niño

6. Magnetic Field
- ❑ Magnetic Forces
- ❑ Earth's Magnetic Field
- ❑ Origin of Magnetic Field
- ❑ Van Allen Radiation Belts
- ❑ The Earth's Magnetosphere
- ❑ Auroras

7. Imaging the Earth
- ❑ Visible Light
- ❑ Earth at Night
- ❑ Radar Imaging
- ❑ IR Imaging
- ❑ Magneto-Gravitational
- ❑ Surface Temperature Maps

Progress Checklist / The Moon

1. Overview of the Moon
- ❑ General Properties
- ❑ Surface Features
- ❑ Interior of the Moon
- ❑ Atmosphere of the Moon
- ❑ Geology of the Lunar Surface
- ❑ Chemistry of the Lunar Rocks

2. Tidal Coupling
- ❑ Tidal Forces
- ❑ Tides Induced
- ❑ Gravitational Locking
- ❑ Consequences of Tidal Forces

3. History of the Moon
- ❑ Initial Formation
- ❑ Melting of the Surface
- ❑ Meteor Bombardment
- ❑ Volcanism

4. Formation of the Moon
- ❑ Competing Theories
- ❑ The Capture Hypothesis
- ❑ The Fission Hypothesis
- ❑ The Co-Accretion Hypothesis
- ❑ The Large Impact Hypothesis
- ❑ Constraints from the Data

Keywords / The Earth

geology
seismic waves
P-waves
S-waves
longitudinal
transverse
refraction
inner core
outer core
mantle
crust
lithosphere
asthenosphere
differentiation
plate tectonics
continental drift
Laurasia
Gondwanaland
Pangaea
radioactive decay
convection current
Mid-Atlantic Ridge
sea floor spreading
San Andreas Fault
oxidizing atmosphere
reducing atmosphere
troposphere
stratosphere
mesosphere
ionosphere
ozone layer
tropopause
solar wind
weather
climate
Coriolis force
cyclone
anticyclone
magnetic field
Van Allen radiation belts
aurora
dynamo effect
magnetosphere
bow shock

Keywords / The Moon

Apollo missions
synodic period
Near Side
Far Side
asymmetric
Maria, mare
Highlands
impact crater
crater density
moonquakes
crust
mantle
core
regolith
ejecta blanket
sedimentary rock
igneous rock
basalt
differential forces
tidal coupling
Roche Limit
gravitational locking
angular momentum
Condensation hypothesis
planetesimal
tidal forces
Capture hypothesis
volatile elements
radioactive dating
Imbrium Basin
volcanism
Fission hypothesis
refractory elements
anorthosite
breccia
refractory elements
volatile elements
radioactive dating
Imbrium Basin
volcanism
Fission hypothesis
Capture hypothesis
planetesimal
tidal forces
differential forces
tidal coupling
Roche Limit
gravitational locking
angular momentum

Exercise 6-1: Introductory Narrative

We learn about the interior of the Earth by studying 1) ____________ waves. There are two types — pressure waves in which particles move along the direction of the wave and 2) ____________ waves (which cannot travel through liquids) where the motion is perpendicular to the direction of the wave. The Earth's present structure came about through 3) ____________, the process whereby more dense materials sink and lighter materials rise to the surface. From the center to the surface, the layers of the Earth's interior are the solid inner core, the liquid outer core, the 4) __________, the aesthenosphere, and the 5) _________. The production of the Earth's magnetic field by currents in the liquid outer core is known as the 6) ____________________. The drifting of the continents is known as 7) ________________. The Earth's atmosphere originally contained mostly hydrogen and helium like the solar nebula. These elements were quickly lost to space and replaced by a secondary atmosphere composed mainly of 8) _________________. This gas was ultimately absorbed by the Earth's 9) ______________. The atmosphere of the earth today is mainly 10) ______________ but is viewed as very unique among the planets because of the large amount of 11) ____________. Three of these atoms may combine to form 12) ____________, which protects the surface of the Earth from harmful ultraviolet radiation. Complex weather patterns exist in the Earth's atmosphere due to the interaction of solar heating and 13) ____________ forces.

The Moon is the nearest extraterrestrial object to us and the only one that we have explored directly. Consequently, we know quite a bit about it. The Moon has a diameter about 14) ______ the diameter of the Earth and a mass of about 15) ___________ the mass of the Earth. The plane of the Moon's orbit is inclined by 16) _________ to the plane of the ecliptic, so eclipses don't occur every month. The period of 17) ____________ is equal to its period of revolution due to the long-term effects of tidal forces.

Most of the rocks on the Moon are 18) _________, meaning "fire-formed rocks." When we look at the Moon we can see bright areas, which are called 19) ____________. This type of terrain is composed of 20) ________________, which is formed by lava that cools very slowly. We can also see darker areas known as maria. This type of terrain is composed of 21) ___________, which are formed by lava that cools quite rapidly. Very few maria are found on the 22) ______________ of the Moon. The maria rocks are the darker and denser of the two types since they contain more iron. Both are covered with a fine layer of powdered rock known as 23) ______________ from many years of meteoric bombardment.

The composition of the Moon is similar to that of the Earth but has much lower 24) ___________, since it has very small amounts of heavy metals. It also has higher amounts of 25) ___________ materials, such as titanium, and lower amounts of volatile materials, such as water. These differences in composition make the 26) __________________ hypothesis the most likely theory for explaining the origin of the Moon. Thus, the material that forms the Moon would have come from the Earth's crust and mantle and have fewer of the heavy materials that are found in the Earth's core.

Exercise 6-2: True/False Questions

T / F 1. Seismic waves are produced naturally by earthquakes, volcanoes, and impacts.

T / F 2. Shear waves can travel straight through the center of the Earth.

T / F 3. Radioactive decay was one of the major sources of heat which allowed the Earth to differentiate.

T / F 4. Earthquakes occur primarily in the center of crustal plates.

T / F 5. The movement of crustal plates is due to convection currents in the asthenosphere.

T / F 6. Chemists refer to the Earth's atmosphere as a reducing atmosphere.

T / F 7. The Earth's ozone layer provides protection from infrared radiation from the sun.

T / F 8. Water has a relatively large heat capacity compared to other substances.

T / F 9. In the United States' mainland, the Coriolis Force causes winds blowing primarily from the west.

T / F 10. A projectile fired from the equator toward the south pole would be deflected to the west by the Coriolis Force.

T / F 11. The northern lights are caused by the Coriolis Force.

T / F 12. Radar can penetrate the clouds of Venus and allow us to map the surface.

T / F 13. Infrared radiation detected by satellites allows us to map the temperature of the surface of the Earth.

T / F 14. We can see more than 50% of the lunar surface due to librations.

T / F 15. The large amount of seismic activity that occurs on the Moon is due to plate tectonic activity.

T / F 16. The Moon was unable to retain an atmosphere because of its very low mass.

T / F 17. The most common type of rocks found on the Moon are sedimentary.

T / F 18. Tidal forces occur because the Earth is pulling differently on different parts of the Moon due to the one over r^2 dependence of Newton's Law of Gravitation.

T / F 19. The tidal locking between the Earth and the Moon is unique in our solar system.

T / F 20. The interior of the Moon is heated by tidal forces from the Earth.

T / F 21. The meteor bombardment in our Solar System is much larger today than it was several billion years ago.

T / F 22. Cratering on the Earth has been comparable to that on the Moon but the evidence is quickly erased by plate tectonic activity and erosion.

T / F 23. Lunar volcanoes tend to occur in long linear chains just like volcanoes on the Earth.

T / F 24. The Moon continues to be geologically active today.

T / F 25. The capture hypothesis has fallen out of favor since it is unlikely that the Earth and Moon would have such similar compositions if the Moon had been formed elsewhere.

T / F 26. The fission hypothesis has fallen out of favor since it is unlikely that the Earth and Moon would have such different compositions if the Moon had been part of the same body.

T / F 27. The best theory for lunar formation suggests that the Moon was formed from a ring of material knocked from the Earth in a large collision.

Unit 7
Mercury, Venus, and Mars

Chapter Objectives

The planet Mercury is the object in our Solar System with a surface most closely resembling the maria and highlands of our Moon. Venus is the planet that most closely resembles Earth in size and overall composition. These two inner planets teach us to look beyond first impressions. Mercury is not at all like our Moon inside and Venus is dramatically different from Earth in surface conditions. This chapter will reveal the surface features, interior, atmospheric composition, and geologic history of these two quite different planets. We will see how spacecraft and radar imaging increased our knowledge of both Mercury and Venus. This chapter will also explore how the inhospitable conditions found on Venus may serve as a warning for us on Earth.

Mars is the second closest planet to Earth, and the one that has always generated the most speculation concerning the possibility of life on its surface. Understanding how Mars resembles Earth and the many ways that it differs from Earth will be the primary goal of this chapter. The results of spacecraft studies of Mars will be delineated and the resultant knowledge of the history of Mars that has emerged will be discussed. The astrobiology experiments performed by the Viking landers and the puzzling results of those experiments will be illustrated. The continuing scientific controversy surrounding the meteorites from Mars, which *may* show evidence of Martian fossil lifeforms, will be discussed.

Progress Checklist / Mercury and Venus

1. The Planet Mercury

- ❑ Planetary Properties
- ❑ Surface Features
- ❑ Interior
- ❑ Atmosphere
- ❑ Magnetic Field
- ❑ A History of Mercury

2. The Planet Venus

- ❑ Planetary Properties
- ❑ Surface Features
- ❑ Interior and Magnetic Field
- ❑ Atmosphere of Venus
- ❑ The Runaway Greenhouse
- ❑ History of Venus

Progress Checklist / Mars

1. Overview of Mars
- ❑ Planetary Properties
- ❑ Surface Features
- ❑ The Martian Interior
- ❑ Atmosphere
- ❑ The Viking Missions
- ❑ History

2. Martian Moons
- ❑ Overview of the Moons
- ❑ Phobos
- ❑ Deimos
- ❑ Origin of Small Moons

Keywords / Mercury and Venus

tidal locking
maximum elongation
Mariner 10
intercrater plains
differentiation
scarps
Caloris Basin
maximal cratering density
Barringer Crater
retrograde rotation
Venera
Magellan
Ishtar Terra
Aphrodite Terra
tectonic activity
carbon dioxide
runaway greenhouse effect
sulfuric acid clouds
nitrogen
Maxwell Montes
rift valley
infrared radiation
synthetic aperture radar

Keywords / Mars

polar ice caps
dry ice
shield volcanoes
Olympus Mons
Valles Marineris
plate tectonics
dust storms
water erosion
wind erosion
differentiated
Mariner 4, 6, 7
Mariner 9
metabolism
mass spectrometer
gas exchange experiment
labeled release experiment
pyrolytic release experiment
peroxides
photosynthesis
meteorite ALH84001
Pathfinder
Mars Global Surveyor
Phobos
Deimos

Exercise 7-1: Introductory Narrative

Mercury is about 1) _________ from the sun on average and is the innermost planet of our Solar System. Because of its nearness to the sun, (it is at most only 28° away from the sun at 2) ___________________), it is very difficult to observe, and many people have never seen it. By using Kepler's third law, one can show that Mercury must have the largest orbital 3) ___________ of all the planets. It orbits the sun in 88 days and completes a rotation in 4) _______ days, two-thirds of the time for a revolution. Since it rotates so slowly, there is a vast temperature difference between the side facing the sun (450°C) and the side facing away from the sun (-180°C). The slow rotation is likely the cause of Mercury having a very small 5) ___________________.

The surface history of Mercury is very similar to that of the Moon, especially in the highland areas. However, the lowlands show considerably less 6) _____________ and far fewer lava flows. Another difference is in composition. Mercury shows a relative abundance of 7) ________________ while the moon has a scarcity of these materials.

Venus is the second planet in our Solar System and revolves around the sun at a distance of 8) _______ in a nearly circular orbit. The most interesting characteristic of the motion of Venus is its slow 9) _____________ rotation. It is theorized that this is probably due to a major collision early in the formation of the planet.

The surface of Venus is extremely inhospitable. The temperature is the hottest in our Solar System at about 500°C due to the 10) ____________ effect. The surface pressure is 11) _________ times that of the Earth and the bright clouds of Venus are largely composed of 12) __________________. Thus, the space probes that have landed there have not lasted very long. Most of the information we know about the surface of Venus was obtained from orbit by the 13) _______________ space probe.

Mars is somewhat similar to the Earth in that one day lasts about 24 hours. The orbital tilt of the planet is also very similar to the Earth, giving the two planets similar 14) ____________. However, Mars has much smaller values of radius, mass, and 15) __________ than the Earth.

Mars has an interesting and dynamic appearance. It is known as the 16) _______________ because of the presence of iron compounds (like rust) and clays in its soil. The appearance of Mars changes due to 17) ___________ and the size of the polar caps varies with the seasons. The atmosphere of Mars is very thin but does have clouds.

Mars has many interesting geological features. There are many large shield volcanoes on Mars, the largest of which is known as 18) _______________. The volcanoes are larger on Mars than on the Earth due to the fact that Mars has no 19) __________________. Surface markings originally called 20) ____________ are now known to be the edges of mountain ranges. There is considerably evidence on Mars of water erosion, suggesting that the Martian atmosphere was 21) ______________ in the past, which would allow liquid water to exist.

Evidence of liquid water on the surface of Mars in the past has fostered speculation that possibly life existed there in the past. In 1976, the 22) __________ space probes looked for life by assuming that it would have a similar metabolism to life on the Earth. Although some interesting soil chemistry was discovered, no evidence for life was found.

Exercise 7-2: The Rotation of Mercury and Venus

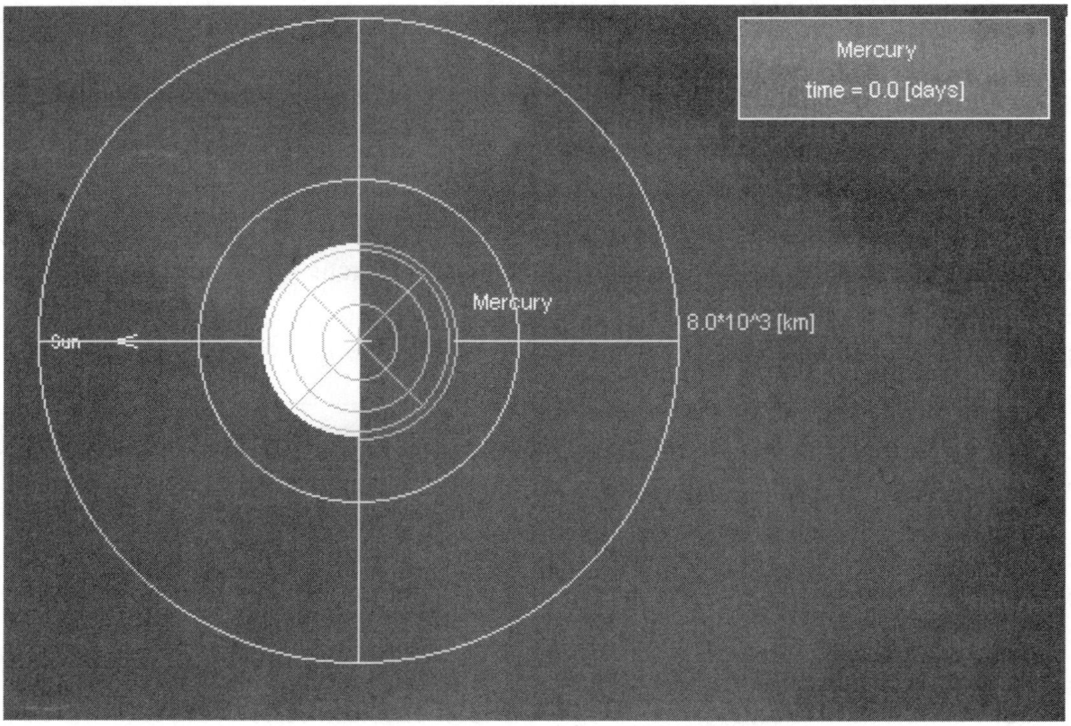

This exercise will make use of (**IC 7.1**). Use the markers button to turn on a yellow arrow that points to the sun and a green meridian. You can use the yellow arrow to time the planet's period of revolution. You can use the green meridian to time the planet's period of rotation.

Start the simulation and time the following:

1) Time for one rotation = ________________
2) Time for one revolution = ________________
3) Time for three rotations = ________________
4) Time for two revolutions = ________________

Question #1: Can you explain the correlation between your answers for questions 3 and 4 above?

____

____

____

Now start up (**IC 7.2**).

Question #2: How do the direction and speed of rotation for Venus compare to Mercury?

____

____

Exercise 7-3: True/False Questions

T / F 1. The density of Mercury is much greater than the density of the Earth.

T / F 2. Mercury has a very large iron core and a very thin mantle compared to the Earth.

T / F 3. There is no detectable atmosphere on Mercury.

T / F 4. The temperature on the side of Mercury facing the sun is the hottest in the Solar System.

T / F 5. Mercury is similar to the Moon in that in that it is very depleted of heavy metals.

T / F 6. Mercury could not retain an atmosphere since its temperature on the side facing the sun is very hot and its escape velocity is relatively low.

T / F 7. Since the clouds of Venus circle the planet in about four days, they must move with very high velocity.

T / F 8. Although the interior of Venus is thought to be similar to that of the Earth, Venus has no magnetic field and no plate tectonic activity.

T / F 9. The atmosphere of Venus is very similar to that of the Earth being mostly nitrogen and oxygen.

T / F 10. There is no evidence for water on Venus and it is thought that it was broken down by UV radiation in the upper atmosphere and the hydrogen escaped into space.

T / F 11. Mars has the highest density among the Terrestrial Planets

T / F 12. One day on Mars is approximately the same length as one day on the Earth.

T / F 13. Since Mars has no plate tectonic activity, the volcanoes of Mars tend to be few in number and extremely large.

T / F 14. The large size of the volcano Olympus Mons suggests that the crust of Mars must be extremely thin for so much lava to have come through it.

T / F 15. Since the orbital tilt of Mars is very small, the planet has no seasons.

T / F 16. Since Mars is farther from the sun than the Earth, it must have had more water on it than the Earth when it formed.

T / F 17. Three out of four basic experiments contained in Viking to search for life returned positive results, which proved that life is present on Mars.

T / F 18. Since the moon Phobos rises in the west and set in the east on Mars, the planet must have retrograde rotation like Venus.

T / F 19. The two moons of Mars are very irregular in shape because they are small and their gravity is insufficient to pull them into a spherical shape.

T / F 20. The two moons of Mars are likely captured asteroids.

T / F 21. Since liquid water cannot exist in the low pressure atmosphere of Mars today, the channels presumably formed by water erosion must have been formed in the past, when the atmospheric pressure was larger.

Unit 8
The Outer Planets

Chapter Objectives

The four Jovian planets are much larger in size and lower in density than the four Terrestrial planets. Jupiter, the largest planet in our solar system, will serve as our introduction to the Gas Giants. Understanding the overall composition and the layered structure of Jupiter, and also appreciating the varied family of satellites orbiting Jupiter are the primary goals of this chapter. The intense atmospheric dynamics of Jupiter's turbulent atmosphere will be described and compared to Earth's prevailing wind patterns and storms. Jupiter's complex magnetic field and the metallic hydrogen layer where it originates will be discussed. The four planetary-sized satellites of Jupiter will be compared and contrasted to each other and to the Terrestrial planets. The smaller satellites will be described, their orbital characteristics illustrated, and the evidence that some may be captured asteroids presented. The basic concept of the ring systems that surround all the Jovian planets will be introduced in discussing Jupiter's thin ring.

One of the most beautiful and intriguing telescopic sights in our Solar System, Saturn has long been known as "The Ringed Planet." No longer unique, Saturn's ring system would still be described as the most spectacular. Describing and understanding Saturn's meteorology, interior, magnetic field, and general properties, as well as its signature rings, are the main objectives of this chapter. Saturn will be compared and contrasted to its larger neighbor Jupiter, which it resembles in many ways. The complex ring system will be explored and the underlying principle of tidal forces will be illustrated. Shepherd satellites, radial spokes, ringlets, and other details of ring structure discovered by the Voyager spacecraft will be discussed. The satellite family of Saturn, especially Titan with its Earth-like atmosphere, will be described.

Uranus, Neptune, and Pluto are the three "discovered" planets, unknown to the ancient astronomers. Uranus and Neptune resemble each other closely and differ significantly from the other two Gas Giants. This chapter will describe the atmospheric dynamics of these two Jovians and show how temperature differences lead to color differences, thus explaining the blue color of Uranus and Neptune. The strange, off center magnetic fields will be illustrated and several hypotheses to explain this odd behavior will be presented. Their possible interior structure and how this may relate to the magnetic field and to the atmospheric dynamics will be explored. The ring systems of Uranus and Neptune and the mechanisms responsible for their formation and maintenance will be described. The smallest and most distant planet, Pluto, and its satellite Charon will be studied using the latest Hubble Space Telescope images.

Progress Checklist / Jupiter

1. Overview of Jupiter
- ❑ Planetary Properties
- ❑ Gas Giant Planets
- ❑ Interior
- ❑ Atmosphere
- ❑ Great Red Spot
- ❑ History

2. The Magnetosphere
- ❑ Magnetosphere
- ❑ Origin
- ❑ Io Plasma Torus
- ❑ Auroras

3. The Galilean Moons
- ❑ Galilean Moons
- ❑ The Moon Io
- ❑ Europa
- ❑ Ganymede
- ❑ Callisto
- ❑ Internal Structure

4. Other Satellites of Jupiter
- ❑ Other Moons
- ❑ The Ring System

Progress Checklist / Saturn

1. Overview of Saturn
- ❑ The Planet Saturn
- ❑ Surface Features
- ❑ Interior
- ❑ Atmosphere
- ❑ Magnetic Field
- ❑ History

2. The Ring System
- ❑ Roche Limit
- ❑ Arrangement and Composition
- ❑ Spokes and Structure
- ❑ Shepherd Moons

3. The Moons of Saturn
- ❑ The Moons of Saturn
- ❑ The Atmosphere of Titan
- ❑ Organic Material on Titan
- ❑ The Other Moons

Progress Checklist / The Outermost Planets

1. The Planet Uranus
- ❑ Planetary Properties
- ❑ Surface Features
- ❑ Interior
- ❑ Atmosphere
- ❑ Magnetic Field
- ❑ History

2. Satellites of Uranus
- ❑ The Moons
- ❑ The Ring System

3. The Planet Neptune
- ❑ Planetary Properties
- ❑ Surface Features
- ❑ Interior
- ❑ Atmosphere
- ❑ Magnetic Field
- ❑ History

4. Satellites of Neptune
- ❑ Moons
- ❑ Triton
- ❑ Ice Volcanoes
- ❑ The Ring System

5. Pluto and Charon
- ❑ Planetary Properties
- ❑ Surface Features
- ❑ Atmosphere
- ❑ Charon

Keywords / Jupiter

Jovian
Galilean
gas giants
Pioneer 10, 11
Voyager I, II
Galileo orbiter
metallic hydrogen
molecular hydrogen
zones
belts
Great Red Spot
anti-cyclone
magnetosphere
plasma sheet
torus
Van Allen Belts
solar wind
bow shock
aurora
Io
Europa
Ganymede
Callisto

Keywords / Saturn

oblate
internal heat source
Great White Spot
Roche Limit
Cassini Division
ringlets
G, F, A, B, and C rings
radial spokes
electrostatic force
shepherd moon
Prometheus
Pandora
Titan
hydrocarbons
Rhea
Mimas
Enceladus
Tethys
Dione
Hyperion
Phoebe

Keywords / The Outermost Planets

methane
ammonia
Miranda
Ariel
Umbriel
Titania
Oberon
petrochemical haze
magnetic dipole
conducting shell
Great Dark Spot
Triton
Nereid
stellar occultation
retrograde revolution
ice volcanoes
Charon

Exercise 8-1: Introductory Narrative

Jupiter has a mass 1) ________ times the mass of the earth and is the largest planet in our solar system. However, Jupiter is primarily composed of hydrogen and helium and thus has a very low 2) ___________ of 1.34 g/cc.

The atmosphere of Jupiter has a striking appearance through a telescope due to its colorful zones and 3) __________. There is also the 400 year old hurricane known as the 4) __________________. As one descends into the Jovian atmosphere the higher pressure gradually converts the hydrogen gas to liquid. However, this happens so gradually that Jupiter has no 5) _________________. It is theorized that in the extremely high pressures of Jupiter's interior, the physical properties of hydrogen are quite different than at lower pressures, and it is known as 6) ___________________________. It is thought that circulating electrical currents in this material are responsible for Jupiter's 7) __________________.

Jupiter has four large moons that were first observed by 8) ____________, since they can be seen from the Earth with small telescopes. The nearest moon to the planet 9) ____________ is very active geologically. It has volcanoes that spew 10) __________ compounds onto the surface and account for the moon's unique appearance. Ions from these volcanoes get trapped in a torus by Jupiter's rapidly rotating 11) ________________. The second large moon, Europa, is extremely smooth and appears to be covered by frozen 12) ____________. It is speculated that life could possibly exist in liquid water underneath the ice. Space probes will likely investigate this in the future. The third moon, 13) ______________, is the largest satellite in our Solar System. It has very different types of terrain, with some sections looking very old and heavily cratered and others formed more recently. The fourth moon, Callisto, is heavily 14) ____________. Thus, the terrain is very old and the moon appears to be geologically dead. The four moons are very different, reflecting their different distances from Jupiter. Jupiter also has many smaller moons that are probably captured 15) _____________, and a faint ring system.

Although Saturn is the second most massive planet in our solar system, it has the smallest 16) _____________. Its composition is very similar to Jupiter's composition, being mainly hydrogen and 17) ______________. In many ways it can be considered a smaller, less colorful version of Jupiter. It has 18) __________________, although none are as substantial as the Great Red Spot on Jupiter. It has a smaller magnetic field than Jupiter, but much higher velocity 19) ____________.

The impressive rings of Saturn are composed of very many small 20) _________. Large gaps such as 21) ____________________ exist between groups of rings. The individual particles making up the rings are mostly 22) ______________. Since the rings are inside Saturn's Roche Limit, this material could never condense due to tidal forces. The rings also contain radial structures called 23) ___________, possibly created by electrostatic repulsion between ring particles. The long term stability of the rings is due to small satellites known as 24) _________________. These satellites orbit near the rings and maintain them through gravitational interactions.

Saturn has 25) ____________ moons, the most impressive of which is Titan. One can see the 26) _____________ of Titan (which is thicker than that of the Earth) as a faint fuzz around the edge of an image of the moon. Clouds are also present composed of drops of liquid nitrogen and 27) ____________. The presence of 28) __________________ on Titan has furthered speculation about the possibility of life evolving there. However, temperatures on Titan are so cold that the chemical reactions necessary for life might be too slow. More information should be gained by the 29) ____________ space probe, which will visit Titan in 2004. Saturn also has six icy moons of medium size, and eleven smaller moons that are probably captured asteroids.

The planets Uranus and Neptune are 30) ____________ planets and thus they are similar to Jupiter and Saturn but have higher concentrations of heavy elements. Uranus is the third largest planet in our solar system, but only the fourth most massive planet since Neptune has a higher 31) __________. Neptune is usually the eighth planet from the sun, but because of Pluto's highly 32) ______________ orbit it is periodically the farthest from the sun. Both Uranus and Neptune have insufficient mass to form liquid metallic hydrogen in their interiors yet both still have strong magnetic fields. Since the fields are considerably inclined with respect to their planet's axes of rotation and are also not centered on the planet there is still much to learn about them.

One of the most interesting characteristics of Uranus is that its axis of rotation is almost in the plane of the 33) ____________. Thus, Uranus has the most extreme 34) ___________ in the solar system. The nebular hypothesis suggests that planets should form with their rotational axes 35) ______________ to the ecliptic plane. The most likely theory to explain this strange orientation involves a 36) ____________.

Neptune has very high 37) ___________ winds and in general has active weather. The very large storm system known as the 38) ________________ resembles Jupiter's Great Red Spot in many ways. This activity is probably due to Neptune having its own internal heat source.

The ninth planet 39) ____________doesn't belong in either the Terrestrial or Jovian group of planets. Although in size it is certainly similar to Terrestrial, its density of 2.1 g/cc is more similar to Jovian. Its orbital 40) _____________ is 17°, by far the largest value. Its moon, 41) _____________ is such a significant fraction of its mass that they really should be considered a binary object. It is different from the other planets in many ways.

Exercise 8-2: Characteristics of Terrestrial and Jovian Planets

Since we have now covered all of the planets in our Solar System, we are now in a position to compare and contrast them. Indicate whether the following statements apply mainly to **Terrestrial Planets**, **Jovian Planets**, **Both**, or **Neither**.

_______________ 1. Have densities comparable to water.

_______________ 2. Have very small orbital inclinations.

_______________ 3. Have relatively large masses.

_______________ 4. Are found close to the sun

_______________ 5. These planets move very rapidly in their orbits

_______________ 6. Revolve around the sun ccw as seen from the NCP

_______________ 7. Are very nearly in the plane of the ecliptic

_______________ 8. Have solid surfaces

_______________ 9. Are found in the outer parts of the solar system

_______________ 10. Have orbital inclinations less than that of Pluto

_______________ 11. Travel around the sun in elliptical orbits

_______________ 12. Are farther from the sun than the asteroid belt

_______________ 13. Have eccentricities larger than that of Pluto

_______________ 14. Have orbital periods that are longer than 2 years

_______________ 15. Have relatively small masses

_______________ 16. Have many moons and rings

_______________ 17. The semi-major axes of their orbits are less than 2 AU

_______________ 18. Rotate in the same direction that they revolve

_______________ 19. Have periods of rotation less than 24 hours

_______________ 20. Have large amounts of hydrogen

Exercise 8-3: True/False Questions

T / F 1. If, upon formation, Jupiter had been 100 times more massive, it would have become a star.

T / F 2. Jupiter's weather is very different from the Earth's weather in that the sun has considerably less influence and the Coriolis force is much larger.

T / F 3. Jupiter's extremely rapid rotation causes it to be fatter at the poles than at the equator.

T / F 4. The temperatures at Jupiter's poles are very similar to the temperatures at Jupiter's equator.

T / F 5. Auroras on Jupiter are very similar to the Aurora Borealis on the Earth, being caused by charged particles form the solar wind interacting with the atmosphere.

T / F 6. Jupiter has a rocky core that is probably similar in composition to the terrestrial planets.

T / F 7. The order of Jupiter's large moons, in increasing distance from the planet, can be remembered with the mnemonic "I Eat Green Carrots."

T / F 8. Jupiter's moon Callisto has active volcanoes.

T / F 9. The volcanic activity of the Galilean Satellites is caused by the tidal forces from Jupiter.

T / F 10. The fact that Callisto is not differentiated suggests that it cooled very slowly after formation.

T / F 11. The small size and irregular shapes of Jupiter's 12 smaller moons suggests that they formed only out of lower density materials.

T / F 12. Particles making up the ring system around Jupiter are kept in place by small moons known as "shepherd moons."

T / F 13. Saturn's extremely rapid rotation flattens it at the poles and makes it the most oblate planet.

T / F 14. Saturn's bands are less colorful than Jupiter's bands because of its higher temperature.

T / F 15. Saturn is very unique in that it is the only planet that has rings.

T / F 16. Space probes traveling to Jovian planets normally require gravity assists from other planets to get there in a reasonable amount of time.

T / F 17. Chemical reactions typically take place more rapidly in colder temperatures, where the particles are moving slowly.

T / F 18. Saturn and Jupiter became large planets because, during their formation, they quickly became sufficiently massive to directly capture material from the solar nebula.

T / F 19. The Roche Limit is the distance from a planet inside of which an object will be ripped apart by tidal forces.

T / F 20. Many unusual structures exist in Saturn's rings, such as spokes, asymmetries, and braids.

T / F 21. The rings of Saturn were likely formed at the same time as the planet.

T / F 22. Titan is thought of as a possible location in which life could develop in our Solar System, because it has an atmosphere and hydrocarbons.

T / F 23. Titan may be covered with liquid oceans of water.

T / F 24. Saturn's moons as a group are comparable with Jupiter's moons.

T / F 25. Uranus was the first planet discovered in modern times.

T / F 26. Uranus and Neptune appear bluer than the other Jovian planet because of the larger percentage of methane they have in their upper atmospheres.

T / F 27. One theory that explains the magnetic field of Uranus is that an icy slush of ammonia conducts electricity near the surface.

T / F 28. Neptune is the only Jovian planet without an internal heat source.

T / F 29. The varied terrain of the moon Miranda can be explained by the theory that the moon was broken into pieces by a collision and then reformed from the resulting debris.

T / F 30. The small moons of the Jovian planets typically have very large albedos.

T / F 31. The rings of Uranus were discovered by the Voyager space probe.

T / F 32. Neptune has the largest density of any of the Jovian planets.

T / F 33. Uranus' moon Miranda shows tremendous geological activity consistent with being broken apart and then reforming.

T / F 34. Neptune's moon Triton has active geysers.

T / F 35. Neptune and Uranus have lower densities than Jupiter and Saturn because they weren't sufficiently massive to retain all of their hydrogen.

T / F 36. Pluto is able to retain an atmosphere because it has a large escape velocity.

T / F 37. At this very instant, Neptune is the farthest planet from the sun.

T / F 38. Pluto is most likely an escaped moon of Neptune.

Conceptual Map 3
Jovian and Terrestrial Planets

In this assignment we will complete a graphical diagram of the planets in our Solar System. The two-page format of the Conceptual Map is ideally suited to describing the planets, since they nicely break into two groups across the page division. Since the planet Pluto doesn't fit particularly well into either group, it has been excluded. The purpose behind this assignment is to organize the characteristics of the Jovian and Terrestrial planets. We want to detail the common characteristics of each group of planets and then elaborate on a few characteristics that uniquely describe each planet.

At the top of each page is a large box for summarizing the characteristics of the two groups. For example, one characteristic of the terrestrial planets could be "high density." The smaller ellipses near the bottom of the page are for detailing the specific characteristics of each planet. For example, important distinguishing characteristics of Uranus are the large axial tilt and the lack of an internal energy source.

Characteristics of Terrestrial Planets

Characteristics of Jovian Planets

-
-
-
-
-
-

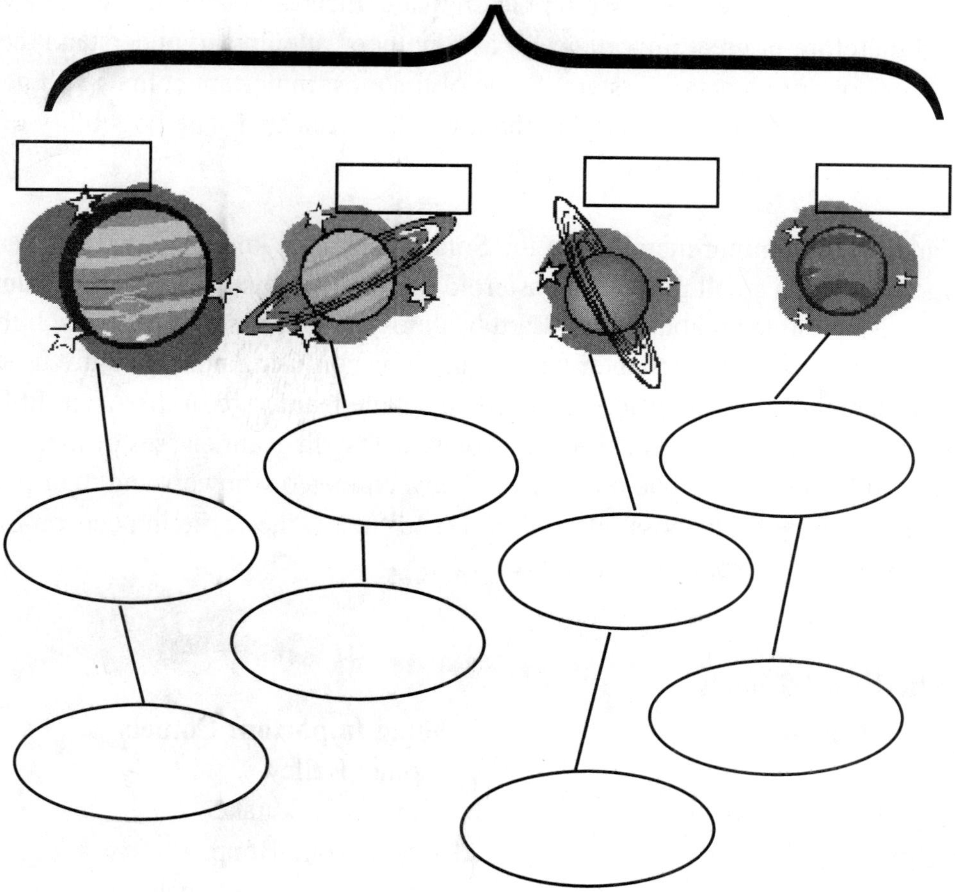

Unit 9
Comets, Asteroids, and Meteorites

Chapter Objectives

The comets are, on average, the smallest of the minor members of the Solar System; but an individual comet can put on a spectacular display in our sky and become for a few weeks the largest object in the Solar System. Comets are composed of dusty ice which vaporizes if the comet comes close to the Sun, giving rise to a long tail which points away from the Sun. This chapter will investigate the origin of these comets and the information gained by telescopic and spacecraft studies of recent comets. We will learn that comets are primordial Solar System material, and therefore of great importance in astronomers' attempts to understand the origins and early history of our planetary system. Some of the most important comets will be described and the data learned by studying them will be delineated. The possibility of Earth colliding with a comet will also be discussed.

The rocky and metallic minor members of the Solar System are known as asteroids and meteoroids. A meteoroid is a small piece of an asteroid or comet and is called a meteor during the brief time that it is visible and burning in Earth's atmosphere. If it survives to reach the surface, it is now a meteorite. This chapter will explain this confusing nomenclature used to describe these samples of solar system debris. The detailed structure of the Asteroid Belt will be described and the reason for its existence explored. The different classes of meteorites and their origins will be discussed. The relationship between meteor showers and their parent comets will be illustrated. The possibility of past (and future) meteorite impacts on Earth and the resultant extinction of species will be discussed.

Progress Checklist / Comets

1. Overview of Comets

❑ Dirty Snowballs
❑ Cometary Observations
❑ Comet Orbits
❑ The Oort Cloud
❑ Head and Coma
❑ The Tail

2. Some Important Comets

❑ Comet Halley
❑ Comet Hyakutake
❑ Comet Hale-Bopp
❑ Collisions with Comets

Progress Checklist / Asteroids and Meteors

1. Overview of Asteroids

- ❑ Asteroid Belt
- ❑ Kirkwood Gaps
- ❑ Gaspra, Ida, and Dactyl
- ❑ Collisions between Asteroids
- ❑ Origins of Asteroids
- ❑ Earth-Crossing Asteroids

2. Meteors

- ❑ Meteor Showers
- ❑ Radiants
- ❑ Common Meteor Showers
- ❑ The Leonid Meteor Shower
- ❑ Fireballs and Bollides
- ❑ Parent Comets

3. Meteorites

- ❑ Classification of Meteorites
- ❑ Finds and Falls
- ❑ Antarctic Meteorites
- ❑ Origin of Meteorites
- ❑ Meteorites from Mars?
- ❑ A Piece of the Asteroid Vesta?

4. The KT Event and Impacts

- ❑ Terrestrial Impact Craters
- ❑ Meteorite Impact Velocities
- ❑ Frequency of Impacts
- ❑ The KT Extinction Event
- ❑ The KT Crater
- ❑ The Asteroids & the Dinosaurs

Keywords / Comets

long elliptical orbit
parabolic orbit
short period comets
long period comets
Oort comet cloud
Kuiper Belt
Chiron
nucleus
coma
ion tail (plasma tail)
dust tail
cometary jets
Comet Halley
Comet Hyakutake
Comet Hale-Bopp
Tunguska Event
Comet Shoemaker-Levy 9

Keywords / Asteroids and Meteors

minor planets
Asteroid Belt
Ceres
Pallas
Vesta
Gaspra
Ida
Kirkwood Gaps
Trojan asteroids
Lagrange points
planetesimals
resonance
Asteroid 1995 CR
Apollo asteroids
meteoroid
meteor
meteor shower
radiant
Perseids
Leonides
Quadrantids
Orionids
Geminids
sporadic rate
fireball
bolide
meteorites
micrometeorites
chondrites (stony meteorites)
iron meteorite
carbonaceous chondrites
pyroxene
Barringer meteor crater
Cretaceous-Tertiary extinction
iridium
Chicxulub relic crater

Exercise 9-1: Introductory Narrative

Comets are best described as 1) ____________________, since they represent material from the primordial solar nebula. Most well-known comets have only slightly elliptical orbits. They tend to stay inside the Solar System and are known as 2) _______________ comets. They are thought to originate in a collection of icy bodies found near the Jovian planets known as the 3) ___________________. Long period comets have highly elliptical orbits. They originate in a large cloud of cometary material that surrounds our Solar System known as the 4) _________________.

At the center of the head of a comet is the 5) _____________, which is typically only a couple of kilometers across and has a density much less than that of water. A comet is heated as it approaches the sun, and gas and dust are ejected from the nucleus and form the 6) ______________, which can be a million kilometers in diameter. As the comet moves, the gas and dust left behind form a tail that can stretch for an AU. Comets often have two tails; a 7) ________ tail that always points directly away from the sun and a 8) __________ tail, which is curved slightly by the solar wind but also reflects the direction from which the comet came. Since comets lose much of their volatile material each time they pass the sun, they can only live through a small number of orbits.

Comets have historically been feared, since they were viewed as omens of important future events. The most famous comet is Halley's Comet, which has an orbital period of 9) ______ years. Comets easily visible to the naked eye only occur about once a decade. We have been extremely fortunate in recent years to view the bright comets Hyakutake in 1996 and 10) _____________ in 1997.

Asteroids are high density objects too small to be planets. Most are found in the 11) _____________________ between the orbits of Mars and Jupiter. The largest is almost 1000 km in diameter and there are estimated to be as many as 12) ______________ larger than 1 km. Asteroids are thought to be material that has never condensed into a planet due to the gravitational perturbations of 13) ______________. This effect can be seen in the 14) ____________________, average orbital radii values for which there are very few asteroids.

As the name implies, Earth Crossing Asteroids are asteroids that cross the Earth's orbit and travel into the inner solar system. They are of special interest due to the possibility of a 15) _____________ with the Earth. It is estimated that there are 16) __________ Earth Crossing Asteroids larger than 1 km. In the last decade alone, 17) __________ asteroids have passed inside of the moon's orbit. Terrestrial Impact Craters show that the Earth has been hit in the past by many large asteroids. A large impact crater on the Yucatan Peninsula in Mexico corresponds to an impact that occurred 65 million years ago and could be responsible for the extinction of the 18) _______________.

The Earth regularly encounters smaller objects known as meteroids. When they hit the Earth's atmosphere with velocities between 10 and 70 km/sec they are vaporized from the heat due to friction and we see a flash known as a 19) ____________, or erroneously as a "shooting star." The larger and more dense of these objects survive until they hit the ground and are then known as 20) _________________. On a typical night this happens a few times per hour. In a 21) ________________________, the rate of "shooting stars" can be as high as 100 per hour and the majority come from one point in the sky known as the 22) ________________. This occurs when the earth is passing through debris left behind by a 23) ________________.

Exercise 9-2: True/False Questions

T / F 1. We can learn a lot about the early Solar System from studying comets.

T / F 2. Astronomers try to discover comets by looking for objects that move with respect to background stars from one night to the next.

T / F 3. Long period comets likely originate in the Kuiper Belt.

T / F 4. A comet nucleus is probably only a few AU in diameter.

T / F 5. The density of a comet nucleus is comparable with that of a terrestrial planet.

T / F 6. The dust tail of a comet always points directly away from the sun due to radiation pressure.

T / F 7. The orbit of Comet Halley is nearly circular.

T / F 8. Jets are commonly seen emanating from the nucleus of a comet.

T / F 9. The Tunguska Event of 1908 was likely the collision of a comet or small asteroid with the Earth.

T / F 10. Comet Shoemaker-Levy was broken into many pieces by the tidal forces of Jupiter.

T / F 11. The gravitational perturbations of Jupiter likely keep the asteroids from condensing into larger bodies.

T / F 12. Non-linear systems are systems in which the response is proportional to the stimulus.

T / F 13. Asteroids are likely the remnants of a planet destroyed in a massive collision.

T / F 14. Although misleading, the term "shooting star" is commonly used to describe meteors.

T / F 15. The location on the earth where the most meteors from a meteor shower land is called the radiant.

T / F 16. Meteor showers occur when the Earth encounters debris left behind by comets as it orbits around the sun.

T / F 17. A meteor that explodes in the Earth's atmosphere is termed a bolide.

T / F 18. The majority of meteorite "falls" are iron meteorites simply because they are the most common.

T / F 19. Meteorites are commonly found in Antarctica because more fall there due to the Earth's magnetic field.

T / F 20. The statement "most meteors come from comets and most meteorites come from asteroids" reflects the higher density of asteroids as compared to comets.

T / F 21. Meteorites have varied composition due to the fact that the planetesimals from which they come differentiate due to heating from radioactivity and are then broken apart in collisions.

T / F 22. Meteorites have been found on the Earth that are believed to be originally from the Moon, Mars, and the asteroid Vesta.

T / F 23. A meteorite found in Antarctica originally came from Mars and shows fossil evidence of early life on Mars

Unit 10
The Sun

Chapter Objectives

Our Sun is the only star we can study up close and in great detail; as such it will serve as the basis of comparison for all the other stars we will investigate. In this chapter we will list the basic properties of our Sun and its overall composition, and we will introduce the Standard Solar Model that attempts to explain these properties. The three directly observable layers of the Sun—the photosphere, chromosphere, and corona—will be described. The complex story of our Sun's changing magnetic field will be illustrated and the field's consequences, including the sunspot cycle and explosive solar flares will be investigated.

Progress Checklist

1. Basic Properties
- ❑ Solar Properties
- ❑ Composition
- ❑ Interior Zones
- ❑ Helioseismology
- ❑ Standard Solar Model
- ❑ Luminosity

2. Photosphere & Spectrum
- ❑ Imaging the Sun
- ❑ Photosphere
- ❑ Opacity and Limb Darkening
- ❑ Frauenhofer Spectrum
- ❑ Granulation
- ❑ Chromosphere

3. Magnetic Field
- ❑ Sunspots
- ❑ Sunspot Cycle
- ❑ Active and Quiet Sun
- ❑ Zeeman Effect and Sunspots
- ❑ The Sun's Magnetic Field
- ❑ Origin of Sunspots

4. The Active Sun
- ❑ Active Regions
- ❑ Prominences
- ❑ Solar Plages
- ❑ Solar Flares

5. Corona and Solar Wind
- ❑ The Corona
- ❑ Corona and Solar Activity
- ❑ Coronal Holes
- ❑ The Solar Wind
- ❑ Coronal Mass Ejections
- ❑ Solar Wind and Earth

Keywords

photosphere
chromosphere
corona
convective zone
radiative zone
core
Frauenhofer lines
hydrogen
helium
helioseismology
velocity field
Standard Solar Model
hydrostatic equilibrium
solar luminosity
solar constant
SOHO
hydrogen-alpha
opaque
transparent
limb darkening
stellar opacity
optical depth
granulation
convection currents
supergranules
spicules
flash spectrum

sunspots
umbra
penumbra
sunspot cycle
sunspot maximum
sunspot minimum
active Sun
quiet Sun
Maunder Minimum
Zeeman effect
magnetic polarity
magnetic cycle
Maunder butterfly diagram
dynamo effect
differential rotation
magnetogram
Babcock dynamo model
active regions
prominences
solar flares
coronal mass ejections
quiescent prominence
eruptive prominence
plages
geomagnetic storms
coronagraph
helmet streamers
solar wind
coronal holes

Exercise 10-1: Introductory Narrative

The Sun is the most studied star, and we believe it to be representative of all stars. The Sun is composed of a mixture of many gases, but it is mainly hydrogen and 1) ________________. It is powered by nuclear reactions in its 2) ____________________. Energy moves to the surface through photons, a process known as radiative transport, and by the circulation of hot material, which is known as 3) ____________________. At all points in the Sun, the inward gravitational forces are balanced by outward gas and radiation pressure in a condition known as 4) ____________________. The 5) ____________________ of the Sun is the total amount of energy leaving the surface each second.

The visible surface of the Sun is called the 6) ____________________. This layer has sufficiently high density to produce enough light to be visible and sufficiently low density to allow this light to escape. Due to convection currents, this layer has a mottled appearance known as 7) ____________________. The region immediately above this layer is the 8) ____________________. Because it is very faint, it is commonly observed only during solar eclipses. The outer atmosphere of the Sun is called the 9) ____________________. It extends out into space many times the radius of the Sun and is much hotter than the surface of the Sun for reasons that are not well understood.

Many energetic and violent phenomena occur on the Sun and are indicative of periods of high solar activity. All of these phenomena appear to be related to the Sun's 10) __________ field. 11) ____________________ are regions on the surface of the Sun that appear dark because they are cooler than surrounding areas. A prominence is a stream of material shot out from the surface of the Sun that often falls backward forming a loop. More energetic eruptions from the surface are called solar 12) ____________________. These phenomena affect the Earth by increasing the flow of charged particles from the Sun known as the 13) ____________________.

Exercise 10-2: True/False Questions

T / F 1. The composition of the Sun is similar to that of the Universe as a whole.

T / F 2. Astronomers know the depth of the convective zone in the Sun from helioseismology data.

T / F 3. The solar constant is the total amount of energy passing each second through the surface of a sphere surrounding the Sun with a radius of 1 AU.

T / F 4. The most important energy transport mechanism in the inner parts of the Sun is convection.

T / F 5. If one can see only a short distance through a fog, its opacity must be large.

T / F 6. The edge of the Sun appears brighter than the center because we are seeing light produced from hotter, deeper layers of the photosphere.

T / F 7. The chromosphere contains spikes of gas called spicules that may represent energy moving from the surface to the corona.

T / F 8. The Sun rotates more rapidly at the equator than it does near the poles.

T / F 9. The frequency of sunspots on the Sun remains fairly constant over time.

T / F 10. The presence of the Zeeman effect indicates that sunspots are magnetic phenomena.

T / F 11. The Babcock model explains the occurrence of sunspots as being due to especially strong convection.

T / F 12. Solar activity fluctuates with an 11-year period in the same way that sunspot counts fluctuate.

T / F 13. Coronal mass ejections increase the number and intensity of auroras on the Earth several days later.

Unit 11
Properties of Stars

Chapter Objectives

In this chapter we learn that our Sun is a very average star that falls in the middle of the stellar range of possible luminosities, surface temperatures, and masses. We will study the laws of nuclear physics in order to understand how stars are able to release tremendous amounts of energy and yet remain so stable during their main sequence lifetimes. We also will use stellar models to describe the energy transport that takes place in the hidden interiors of stars. We will show how astronomers determine the distances to stars and how they trace their motions in our sky. The HR Diagram will be introduced as a graphical summary of some important characteristics of stars.

Over half the stars in our sky are found in pairs (binary stars) or larger groups. Analysis of these binary systems provides us with our best means of determining stellar masses and sizes. In this chapter we will investigate the different types of binary stars and what can be learned from each of them about the physical properties of the stars. We will review Newton's generalizations of Kepler's Laws as tools to determine stellar masses. The mass-luminosity relation that applies to most stars will be developed from these data on stellar masses. The Doppler effect will be applied to a study of binary star systems and also to the discovery of extrasolar planets. We will study the accretion of matter from one star onto another in some binary systems and the relationship between that accretion and novae, supernovae, strong X-ray sources, and black holes.

Progress Checklist / Ordinary Stars

1. Energy Production
- ❑ Mass and Energy
- ❑ Curve of Binding Energy
- ❑ Nuclear Reactions
- ❑ Reaction Rates
- ❑ Temperature and Pressure
- ❑ Energy Windows

2. Stellar Burning Stages
- ❑ Hydrostatic Equilibrium
- ❑ The Proton-Proton Chain
- ❑ The CNO Cycle
- ❑ PP-CNO Competition
- ❑ Triple-Alpha
- ❑ Advanced Burning

3. Energy Transport
- ❑ Energy Transport and Equilibrium
- ❑ Radiative Transport
- ❑ Conduction
- ❑ Convection
- ❑ Neutrino Cooling
- ❑ Cooling Mode Competition

4. Solar Neutrinos
- ❑ Neutrinos
- ❑ Neutrinos Detector
- ❑ The Solar Neutrino Problem
- ❑ Possible Solutions

5. Stellar Distances and Motion
- ❑ Parallax
- ❑ Distance Units
- ❑ Distances to Nearby Stars
- ❑ Proper Motion
- ❑ Space Velocities

- ❏ Motions of the Sun

6. **Stellar Magnitudes**
- ❏ Magnitude Scale
- ❏ Apparent Magnitude
- ❏ Absolute Magnitude
- ❏ Dependence on Wavelength
- ❏ Color Filters
- ❏ Color Indices

7. **Harvard Spectral Sequence**
- ❏ Spectral Sequence
- ❏ Radiation Laws
- ❏ Ionization and Temperature
- ❏ Spectral Sequence Origin
- ❏ Hertzsprung-Russell Diagram

8. **HR Diagram**
- ❏ Main Sequence
- ❏ Giants and Supergiants
- ❏ White Dwarfs
- ❏ Luminosity Classes
- ❏ Spectroscopic Parallax

Progress Checklist / Multiple Stars and Star Clusters

1. **Visual Binaries**
- ❏ Introduction to Visual Binaries
- ❏ Orbits for Binary Stars
- ❏ Proper Motion for Binary Stars
- ❏ The Center of Mass
- ❏ Kepler's Laws and Binary Orbits
- ❏ Tilt of Binary Orbits

2. **Astrometric Binaries**
- ❏ Astrometric Binaries
- ❏ Astrometry and Masses
- ❏ Proper Motion
- ❏ Mass-Luminosity Relation
- ❏ Planets and Binary Star Systems
- ❏ Masses of Binary Star Systems

3. **Spectroscopic Binaries**
- ❏ Spectroscopic Binary Systems
- ❏ Dopper Effect
- ❏ Spectroscopic Binary Animation
- ❏ Spectra
- ❏ Velocity Curves
- ❏ Masses

4. **Eclipsing Binaries**
- ❏ Eclipsing Binaries
- ❏ Light Curves
- ❏ Algol
- ❏ Eclipsing Animation
- ❏ Algol VRML
- ❏ Radii

5. **Accreting Binaries**
- ❏ Accreting Binaries
- ❏ Angular Momentum
- ❏ Accretion Disk
- ❏ Roche Lobes
- ❏ Nova and Supernova
- ❏ Black Holes

6. **Open Clusters**
- ❏ Clusters and Groupings
- ❏ Messier Objects
- ❏ Open Clusters
- ❏ Examples
- ❏ Formation
- ❏ Age and Evolution

7. **Globular Clusters**
- ❏ Overview
- ❏ Examples
- ❏ Formation
- ❏ Ages and Evolution
- ❏ Stellar Populations
- ❏ The Core

Keywords / Ordinary Stars

special theory of relativity
mass-energy conversion
thermonuclear
binding energy
nucleon
proton
neutron
nuclear fission
nuclear fusion
"missing mass"
Coulomb barrier
tunneling
kinetic theory of gases
Maxwellian distribution
Ideal Gas Law

Gamow window
hydrostatic equilibrium
proton-proton chain
Carbon-Nitrogen-Oxygen Cycle
Beta decay
positron
neutrino
deuterium
catalyst
triple-alpha process
radiative transport
conduction
convection
neutrino emission
degradation
random walk
degenerate matter
temperature gradient
solar neutrino problem
neutrino detectors
solar neutrino unit (SNU)
neutrino oscillation
parallax
parallax angle
parsec
light-year
proper motion
space velocity
radial velocity
tangential velocity
solar apex
solar antapex
apparent magnitude
absolute magnitude
color index
Harvard Spectral Sequence
ionization
Hertzsprung-Russell Diagram
main sequence
red giant
white dwarf
supergiant
luminosity class
spectroscopic parallax

Keywords / Multiple Stars and Star Clusters

visual binaries
center of mass
astrometric binary
Kepler's Laws
Kepler's Third Law
tilt angle
proper motion
mass-luminosity relation
extrasolar planets
spectroscopic binaries
Doppler effect
red shift
blue shift
double-line spectroscopic binary
single-line spectroscopic binary
velocity curve
eclipsing binaries
light curve
Algol eclipsing binary
accreting binaries
angular momentum
accretion disk
X-ray binary
Roche lobes
inner Lagrange point
nova
supernova, type I
X-ray burster
black hole
open (galactic) cluster
globular cluster
associations
Messier objects
Pleiades cluster
Hyades cluster
Eagle nebula
Orion nebula
"turnoff" point

Exercise 11-1: Introductory Narrative

Stellar Energy Production

Stars shine by converting mass into 1) ____________________. They do so through 2) ____________________ reactions in their cores. In these reactions two lighter nuclei combine to form a heavier nucleus. Because the heavier nuclei are more 3) ______________, nuclear energy is released in the process. Because the lighter nuclei are both 4) ____________________ charged, they repel each other. This electrical repulsion is known as the 5) ____________________. The nuclei must be traveling very rapidly to get close enough for the strong nuclear force to bind them together before the electrical repulsion pushes them apart. Only in the core of stars is the 6) ____________________ high enough that nuclei have the necessary velocities. In the most common reaction, two hydrogen nuclei are converted into 7) ____________________. In low-mass stars, this occurs primarily by the 8) ____________________ chain and in high-mass stars mainly through the CNO Cycle. The energy is transported from the core of the star to the surface mainly through radiation and 9) ____________________. Although very important in thermal transport on the Earth, conduction plays very little role in stars due to their low densities.

Neutrinos are very low mass (probably massless) particles that are commonly produced in nuclear reactions and interact with matter very weakly. They are an important energy transport mechanism for 10) ____________________ stars. This occurs because more neutrinos are produced in the extremely high core temperatures of these stars and because it is much easier for a neutrino to escape from the very high densities found in these stars than a photon. Neutrino production in the Sun is of special interest because the Sun appears to be producing about 11) ____________________ as many neutrinos as astronomers predict it should on the basis of calculations of nuclear energy reaction rates.

Exercise 11-2: Introductory Narrative

Stellar Parameters

Astronomers are very interested in the intrinsic properties of stars such as distance, velocity, radius, mass, temperature, and composition. These parameters are interrelated in that knowing one of them tells you information about the others.

Distances are determined primarily through the method of 1) ____________________. This can be done only for stars within 100 pc using ground-based observations, but can be extended to about 2000 pc with observations from the 2) ____________________ satellite.

The space velocities of stars have two components that must be evaluated separately. The motion of the star on the celestial sphere is called 3) ____________________ and, if the distance to the star is known, can be used to calculate the tangential velocity. The radial velocity is calculated using the 4) ____________________. One can combine these two components trigonometrically to obtain the space velocity.

The perceived 5) ____________________ of a star is usually specified on a logarithmic system called apparent magnitude. This parameter does not take into account the 6) ________ of the star. The apparent magnitude of the star if it were moved to a distance of 10 pc is known as the 7) ____________________. This parameter describes the intrinsic brightness of the object.

The Harvard Spectral Sequence is used to describe the surface 8) ____________________ of a star. The sequence is denoted by the sequence of letters O, 9) __________, A, F, G, 10) __________, and M. It is determined by looking at the absorption lines in the spectrum of a star. Similar information is conveyed by the 11) ____________________ of a star, which is the difference between the magnitudes through two different filters.

A diagram of absolute magnitude or luminosity versus spectral type or color index is known as a(n) 12) ____________________ Diagram. The location of a star on this diagram is closely related to the star's 13) ____________________.

Exercise 11-3: Introductory Narrative

Binary Stars and Clusters

Most stars are not single stars like our Sun, but instead are part of two-star systems called binaries. Astronomers recognize several different types of binary stars. Systems where we can visually distinguish the two stars in orbit around each other are called 1) ______________ binaries. If only one of the two stars is bright enough to be seen, astronomers may still be able to discern that it is a binary by the wobbles in its proper motion about the center of mass. Binaries that are spotted in this way are known as 2) ______________ binaries. Binaries that are identified through Doppler shifts in their spectral lines are known as 3) ______________ binaries. Graphs of radial velocity versus time are known as 4) ______________ and provide detailed information about the system. Binaries where one star actually passes in front of the other along our line of sight are known as 5) ______________ binaries. These systems must have an angle of inclination *i* near 6) ______________. These systems are normally studied by graphing the variations in brightness as a function of time, which is called a(n) 7) ______________. They afford astronomers one of the few direct ways by which stellar 8) ______________ can be measured. Some binary pairs are close enough that matter from one star forms a(n) 9) ______________ around the second star as the matter spirals into the second star.

Stars are also found in larger groups known as clusters. 10) ______________ clusters contain small numbers of young stars and are found within the disk of our galaxy. The stars they contain are high in metals and are known as population I stars. 11) ______________ clusters contain large numbers of small older stars and are symmetrically distributed about the center of our galaxy. The stars they contain are especially low in "metals" and are known as 12) ______________ stars. This second type of cluster tends to be larger and contains about 1000 times as many stars as the first type.

Exercise 11-4: Using Parallax*

In this exercise we will apply the concepts of parallax to making distance measurements on the Earth. These are commonly employed by surveyors to determine the distance to an inaccessible location such as the distance across a river.

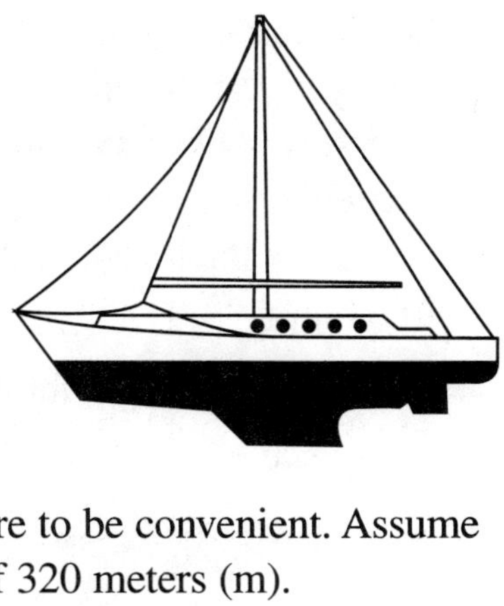

Our goal in this exercise is to determine the distance to a yacht anchored far out from shore in Thunder Lake. A diagram illustrating this scenario can be found on the following page and will be developed in the following paragraphs. The major concept behind parallax is that nearby objects appear to be in different locations relative to more distant objects when the perspective of the observer changes. Thus, we want to look at the position of the yacht from two different locations separated by a distance we call the baseline. The larger the baseline is, the more accurate the technique will be, but we also want the procedure to be convenient. Assume we have paced out or used a measuring tape to define a baseline of 320 meters (m).

We now want to make an angular measurement defining the position of the yacht from each end of the baseline. Surveyors have very accurate tools for making such measurements. They consist of a small spotting telescope attached to a precise protractor mounted on a tripod. Thus, the surveyor can sight on the yacht and measure the angle to the yacht relative to the baseline. Let's assume you have access to such an apparatus. From position A you measure an angle of 81° from the baseline to the yacht and from position B, 72°. You could do this very crudely with just a protractor.

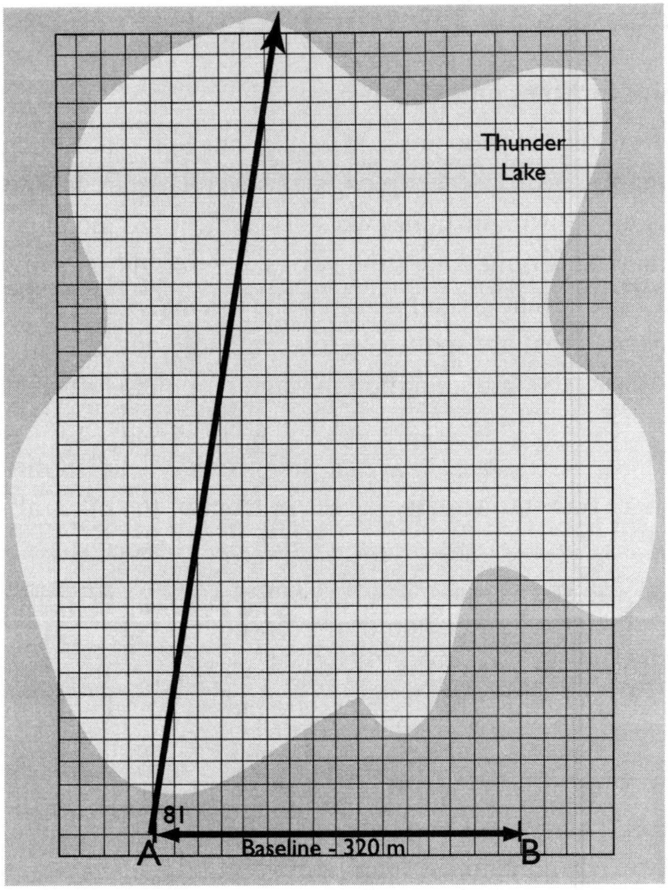

We would like to determine the distance from the yacht to the baseline along a line perpendicular to the baseline. One method of doing this would be to approach the problem mathematically and apply trigonometry. To avoid mathematical complexity, we will take the simpler route of drawing a scaled diagram of the real-world situation on graph paper. In this diagram all distances will be proportional to the their real-world counterparts—only smaller. We first need to determine an appropriate scale so that our diagram will fit on the graph paper. The choice made here was to let the 320 m baseline be 16 blocks on the graph paper so that the scale is 1 block = 20 m. From position A we then draw a line at an angle of 81°. The intersection of this line with the line from position B defines the position of the yacht. We have now completed the diagram found on this page. You should now finish this parallax procedure and determine the distance to the yacht.

1. Draw in the line from position B at an angle of 72° with respect to the baseline. Draw in the yacht at the intersection of the two lines.
2. Draw in a line from the yacht to the baseline that makes an angle of 90° with the baseline.
3. Measure the length of this line:

   Length = ____________________ blocks
4. Convert this length into meters using the scale factor.

   Distance to yacht = ____________________

Exercise 11-5: Using Magnitudes*

Astronomers typically use a magnitude scale to classify stars because of the wide range of brightnesses that they have. We can use the following equation to relate the magnitude difference between two stars to the ratio of their brightnesses:

$$m_2 - m_1 = 2.5 \log\left(\frac{b_1}{b_2}\right)$$

We can reformat this equation as

$$\frac{b_1}{b_2} = (2.5)^{(m_2 - m_1)}$$

Thus, each difference of one magnitude corresponds to a factor of 2.5 in intensity. So a first magnitude star is ($2.5 \times 2.5 \times 2.5$ =) 16 times as bright as a fourth magnitude star. This means that your eye is detecting 16 times as many photons of light.

The magnitude system described above is referred to as apparent magnitude. Stars with lower apparent magnitude appear brighter to us but may not be intrinsically more luminous because we haven't yet taken into account their distances. We do this with absolute magnitude (M). This is the apparent magnitude (m) of a star moved to a distance of 10 parsecs (pc). Thus, we imagine that we can move all of the stars onto a sphere with a radius of 10 pc that surrounds us. With the distance factor effectively removed, the apparent (now absolute) magnitude would allow us to identify the intrinsically more luminous objects. We can relate the apparent and absolute magnitudes through the following equation:

$$m - M = -5 + 5 \log_{10} d$$

where the quantity $m - M$ is known as the distance modulus. Knowing the distance modulus allows one to calculate the distance. The values for both equations have been tabulated to help avoid mathematical complexity.

Magnitude Difference ($m_2 - m_1$)	Brightness Ratio (b_1/b_2)
0	1
1	2.5
2	6.3
3	16
4	40
5	100
6	250
7	630
8	1,600
9	4,000
10	10,000
15	10^6
20	10^8
25	10^{10}

Distance Modulus ($m - M$)	Distance (d in pc)
0	10
1	16
2	25
3	40
4	63
5	100
6	160
7	250
8	400
9	630
10	10^3
15	10^4
20	10^5

Directions: Use the two tables above to answer the following questions concerning the newly created 89th constellation. Note that the apparent magnitudes are given. Consider only labeled stars.

1. The star that would appear brightest to you is ____________________.
2. The star that would appear faintest to you is ____________________.
3. The light from ____________________ is 16 times less intense than the light from δ pistolis.
4. The light from ____________________ is 100 times more intense than the light from κ pistolis.

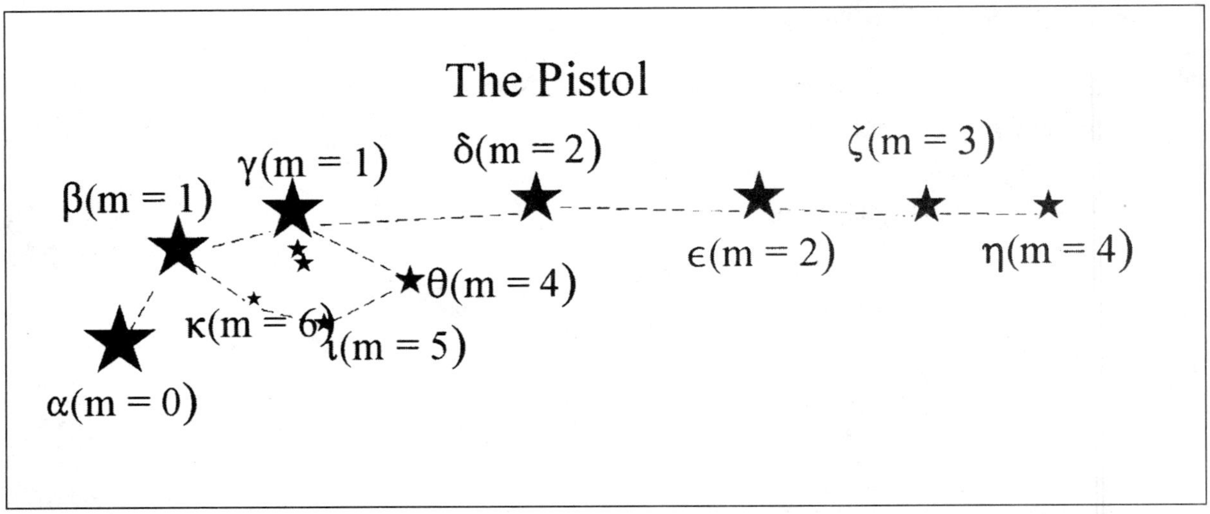

5. The intensity ratio between ζ pistolis and κ pistolis is ______________________.
6. The intensity ratio between β pistolis and ι pistolis is ______________________.
7. η pistolis is 10 parsecs distant; thus, its absolute magnitude M is ______________________.
8. γ pistolis is 40 parsecs distant; thus, its absolute magnitude M is ______________________.
9. The star ϵ pistolis has $M = -4$; thus, it is ______________________ parsecs distant.
10. The star β pistolis has $M = 0$; thus, it is ______________________ parsecs distant.
11. θ pistolis is 140 parsecs distant; thus, its absolute magnitude is ______________________.
12. The star δ pistolis has $M = -7$; thus, its distance modulus is ______________________.

Exercise 11-6: The HR Diagram*

Directions: Indicate the name(s) of the stars from the diagram to the right that correspond to each of the following statements.

1. The hottest star

   ____

2. The coolest star

   ____

3. The most luminous star

   ____

4. The least luminous star

   ____

5. The star with the largest radius

   ____

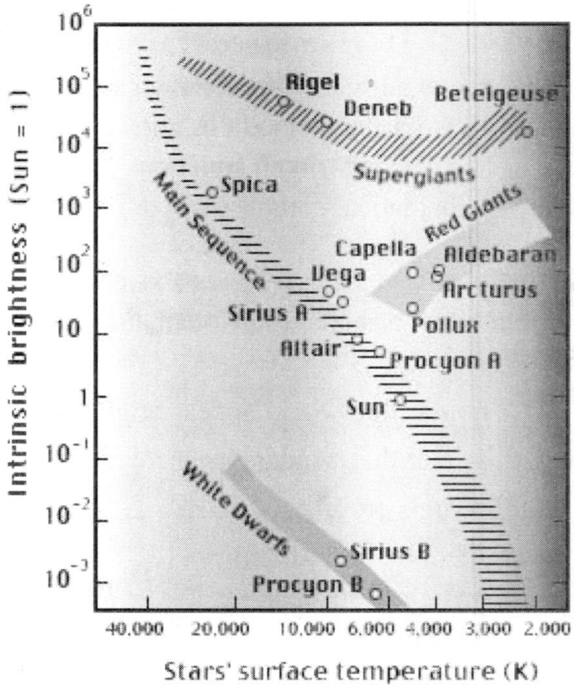

6. The star with the smallest radius

____

7. Main sequence stars

____

8. A spectral type A0 star

____

9. A luminosity class 1a star

____

10. A star with titanium oxide lines in its spectra

____

Exercise 11-7: Spectroscopic Parallax*

In this assignment we will apply the distance determination technique known as spectroscopic parallax. To use it, an astronomer must obtain photometric observations of both the star's apparent magnitude and its spectra. You should keep in mind that even professional astronomers can achieve only 30% accuracy using spectroscopic parallax. This exercise has a wide range of correct answers.

The procedure can be summarized as follows:

1. Use the particular absorption lines present in the spectrum to determine the spectral type from the chart below.
   Example: If your spectrum has strong hydrogen and weak ionized calcium lines (calcium is an ionized metal), you could classify it as approximately A0.

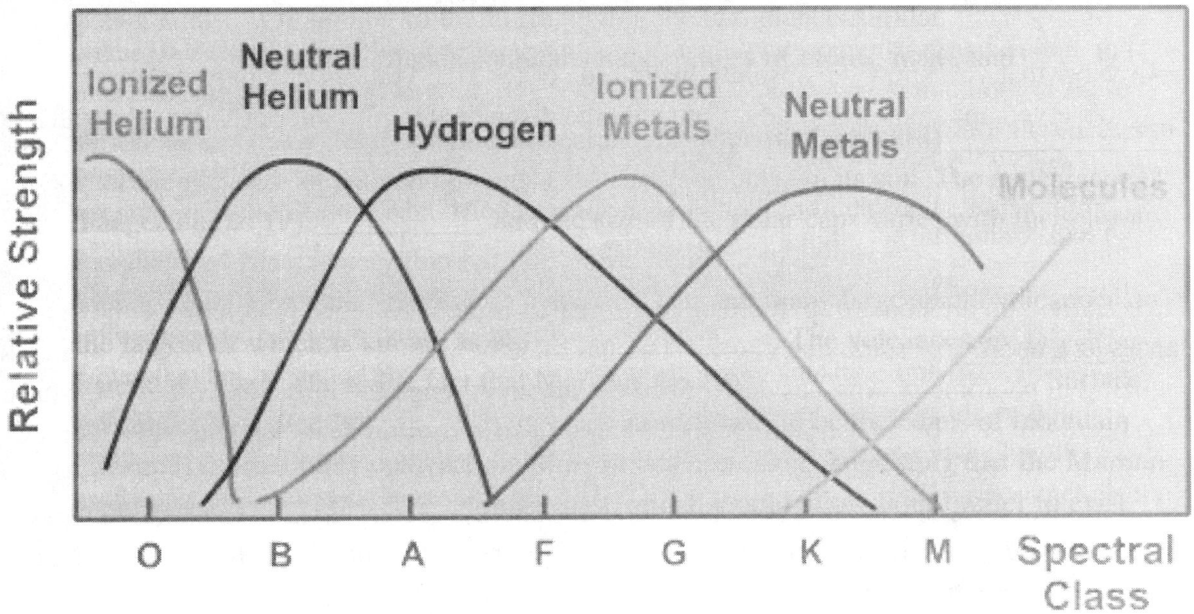

2. Use the thickness of the absorption lines present to determine the luminosity class from the diagram on the next page. Very thick lines will be luminosity class V (main sequence stars), and very thin lines will be luminosity type Ia.

Example: If your spectrum has very thick spectral lines, you could classify it as luminosity class V.

3. We now have uniquely specified the location of the star on the diagram below. Simply find the intersection of the spectral type (a column on the graph) and the luminosity class curve. Once you have noted this position, you can move horizontally to the left on the graph and read off the star's absolute magnitude.
   Example: Find the intersection of spectral type A0 and luminosity class V on the diagram below. Now move horizontally to the left of this point and read off the absolute magnitude, which is about +1.0.

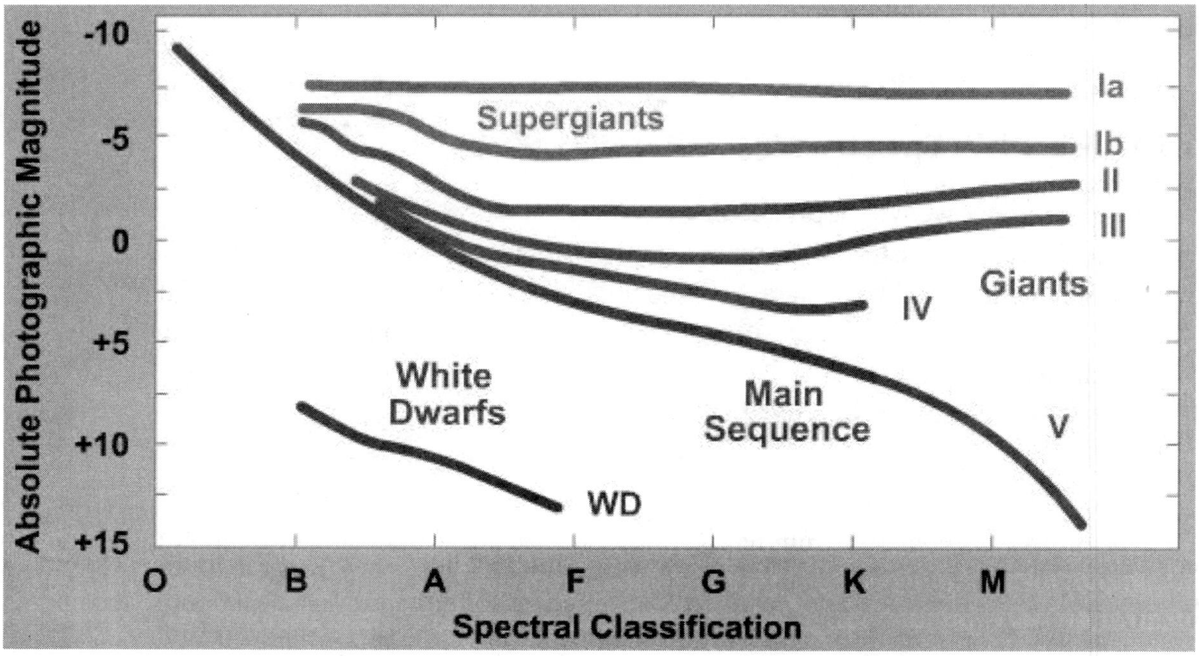

4. Now that you know the absolute magnitude of the star, you can compare it to the observed apparent magnitude. Calculate the distance modulus and look up the distance to the star in the second table of Exercise 18-4.
   Example: If the apparent magnitude is $m = 9$, then the distance modulus is $m - M = 9 - 1 = 8$. Using the second table of Exercise 18-4, we find a distance of 400 pc.

Directions: You are about to participate in a research project designed to determine the distances to a number of stars. Each of the stars has already been observed by an astronomer specializing in photometry who has determined the apparent magnitude m of each star. They have also been observed by an astronomer specializing in spectroscopy who has given you a crude description of the types of absorption lines present in each star's spectra and the thicknesses of those spectral lines. Using the above two diagrams and the second table of Exercise 18-4, estimate the distance to each of the following stars using the technique of spectroscopic parallax. Complete each entry in the table on the following page as you go.

Star 1: $m_V = 9$, strong hydrogen and weak ionized calcium lines, very thick spectral lines

Spectral Type: ____________________ Luminosity Class: ____________________

Absolute Magnitude: ____________________ Distance Modulus: ____________________

Distance (in pc): ____________________

Star 2: $m_V = 4$, strong hydrogen and weak ionized calcium lines, very thin spectral lines

Spectral Type: ____________________ Luminosity Class: ____________________

Absolute Magnitude: ____________________ Distance Modulus: ____________________

Distance (in pc): ____________________

Star 3: $m_V = 6$, very weak hydrogen and very strong neutral iron lines, thin spectral lines

Spectral Type: ____________________ Luminosity Class: ____________________

Absolute Magnitude: ____________________ Distance Modulus: ____________________

Distance (in pc): ____________________

Star 4: $m_V = 7$, weak hydrogen and strong neutral helium lines, very thick spectral lines

Spectral Type: ____________________ Luminosity Class: ____________________

Absolute Magnitude: ____________________ Distance Modulus: ____________________

Distance (in pc): ____________________

Star 5: $m_V = 10$, very weak hydrogen and titanium oxide lines, medium spectral lines

Spectral Type: ____________________ Luminosity Class: ____________________

Absolute Magnitude: ____________________ Distance Modulus: ____________________

Distance (in pc): ____________________

Star 6: $m_V = 8$, medium hydrogen and very strong ionized calcium, thick spectral lines

Spectral Type: ____________________ Luminosity Class: ____________________

Absolute Magnitude: ____________________ Distance Modulus: ____________________

Distance (in pc): ____________________

Exercise 11-8: Eclipsing Binary Light Curves*

Directions: Complete the graphing and questions concerning the light curve of the eclipsing binary star system shown below. Note that star A has a hotter surface temperature than star B.

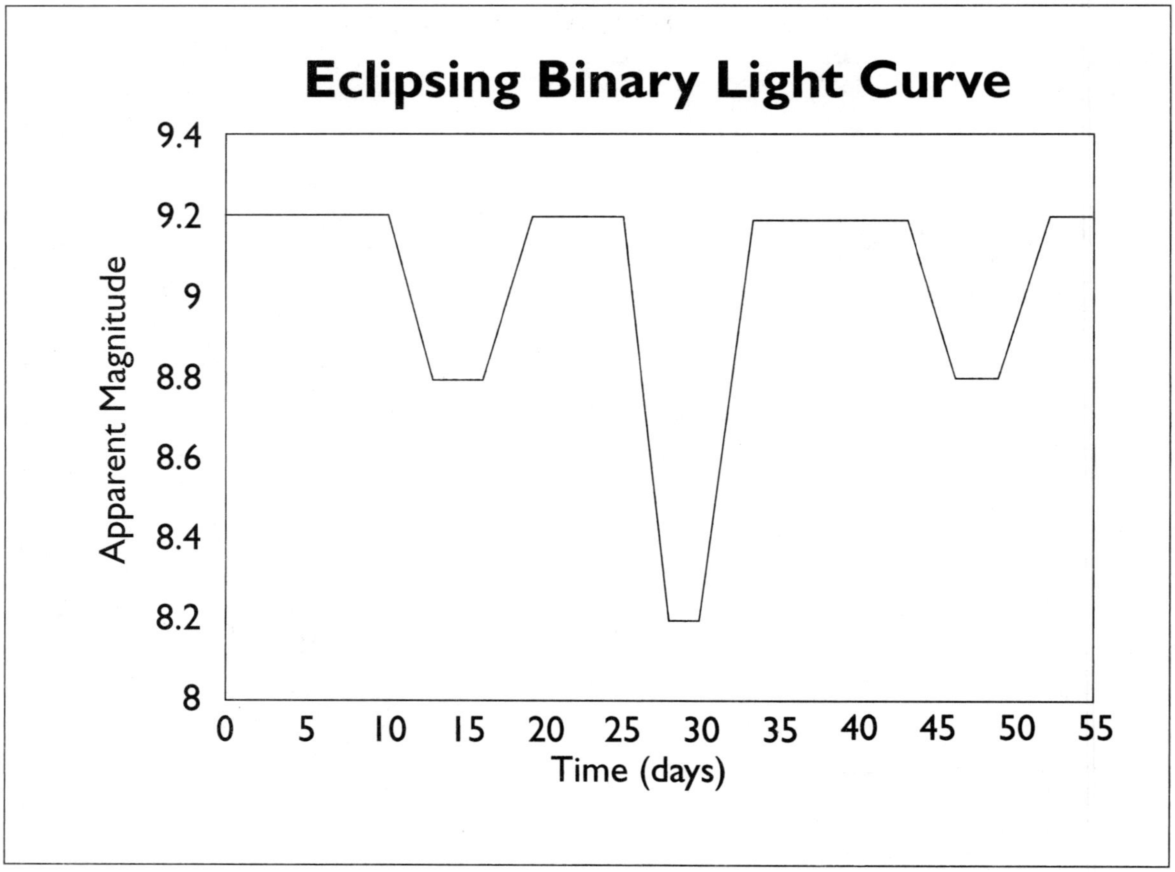

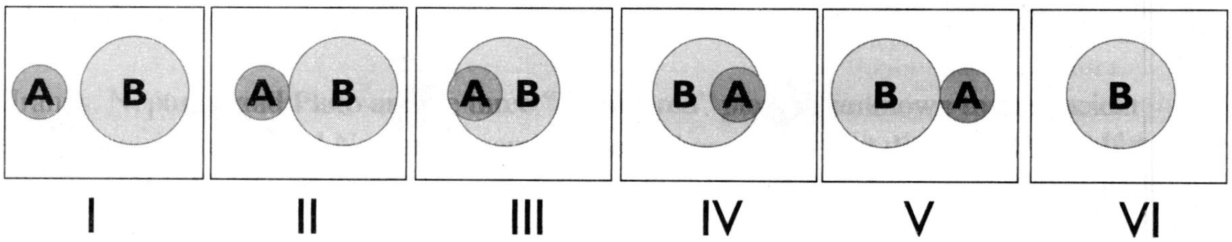

1. The six boxes above all correspond to a certain location on the light curve. Place the roman numeral corresponding to each figure at an appropriate location (time) on the light curve above.
2. The orbital period of the binary is ________________.
3. The depth of the primary eclipse is ________________, and the depth of the secondary eclipse is ________________.

Exercise 11-9: Spectroscopic Binary Stars*

Directions: Complete the graphing and questions concerning the radial velocity curve of the double-line spectroscopic binary shown below.

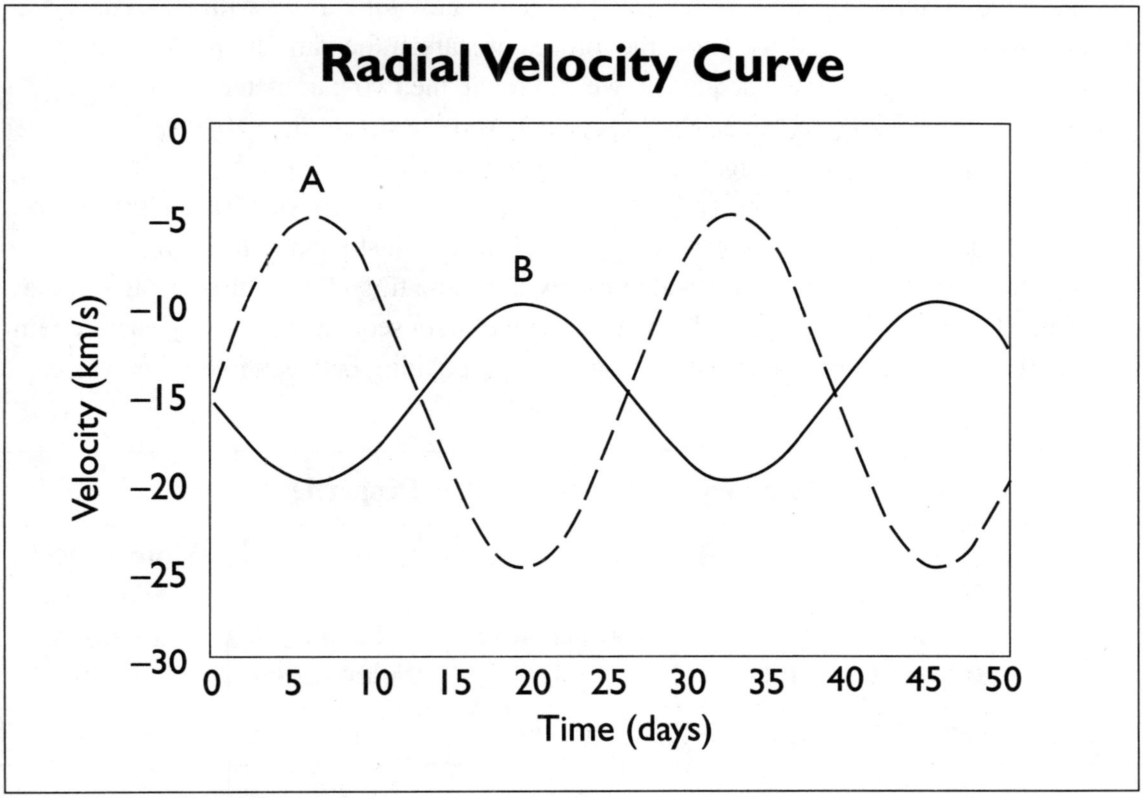

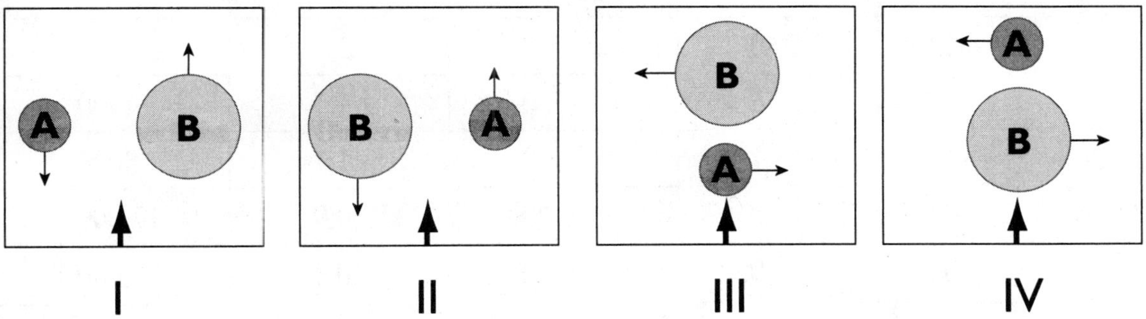

1. The four boxes above all correspond to a certain location on the radial velocity curve. The thick arrow at the bottom of each box represents the line of sight from the Earth, and the thin arrows represent the instantaneous velocities of the stars. Place the roman numeral corresponding to each figure at an appropriate location (time) on the radial velocity curve above.
2. The orbital period of this binary is ____________________.
3. The radial velocity of the center of mass of the system is ____________________.
4. Star ____________________ is the more massive of the two stars.

Exercise 11-10: The Ages of Open Clusters*

In this exercise we will use the turnoff point of a cluster on an HR Diagram to crudely estimate the age of the cluster. The turnoff point is the location on the HR Diagram where stars are starting to leave the main sequence. These stars are running out of fuel in their cores and are starting to move up the red giant branch. Since we know how long a star with a certain mass lives on the main sequence and we know the masses of stars at the various spectral types/surface temperatures along the main sequence, we can relate the two parameters.

The table entitled Main Sequence Spectral Class Properties from the *Properties of Stars/The Hertzsprung-Russell Diagram/Main Sequence* module is shown below. If we can identify the temperature (or spectral type) of the turnoff point on a cluster HR Diagram, we can use the chart to look up the corresponding age of the star just leaving the main sequence. Since all of the stars of the cluster formed at nearly the same time, this is also the age of the cluster. Since the table does not provide the time on the main sequence for every surface temperature, it will be necessary to estimate the age by interpolating between two given values.

Main Sequence Spectral Class Properties

Spectral Class	Mass (Solar Units)	Luminosity (Solar Units)	Temperature (K)	Radius (Solar Units)	Time on Main Sequence (Million Years)
O5	40	400,000	40,000	13	1.0
B0	15	13,000	28,000	4.9	11
A0	3.5	80	10,000	3.0	440
F0	1.7	6.4	7500	1.5	2700
G0	1.1	1.4	6000	1.1	8000
K0	0.8	0.46	5000	0.9	17,000
M0	0.5	0.08	3500	0.8	56,000

Directions: Determine the turnoff point for each cluster shown on the following page. Be careful not to be fooled by any blue stragglers. Draw a vertical line at the temperature corresponding to the turnoff point and read off the temperature. Look up the temperature in the chart above and use it to estimate the age of the cluster in millions of years. Write in your estimated age below each cluster HR Diagram and circle any stars that you determine to be blue stragglers.

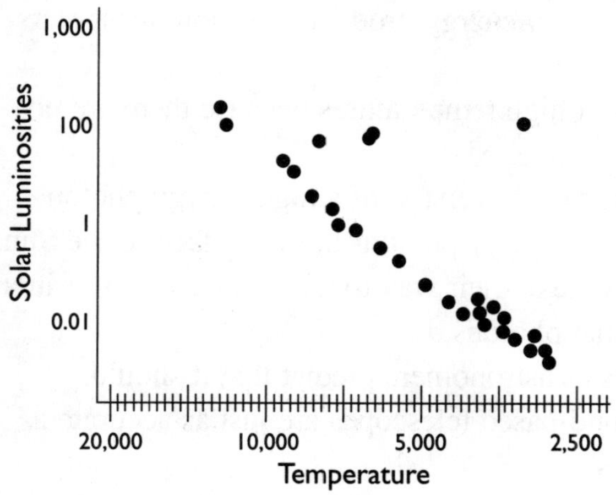

Age: ______________________

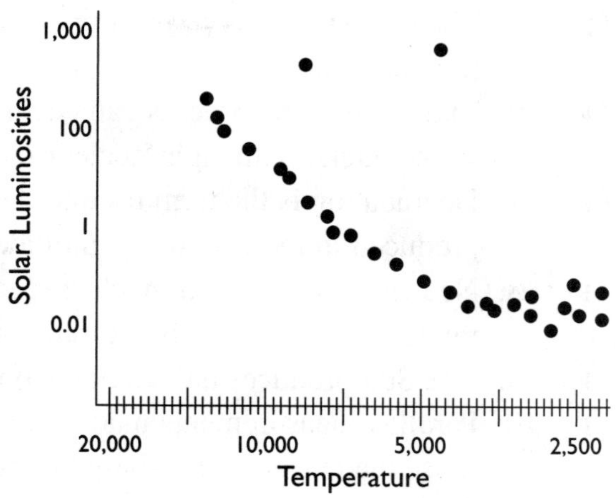

Age: ______________________

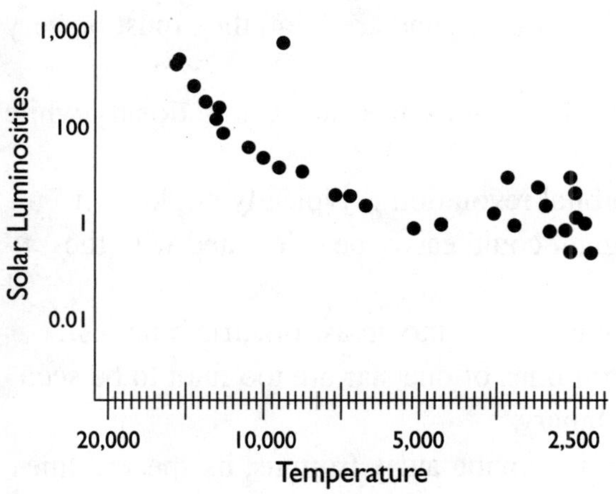

Age: ______________________

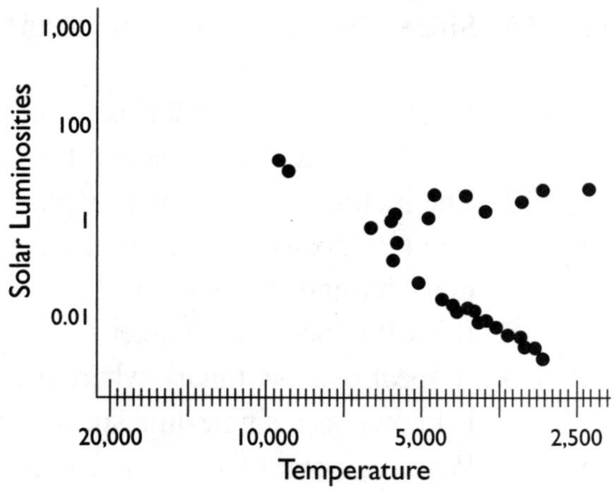

Age: ______________________

Exercise 11-11: True/False Questions

T / F 1. Energy production in a star comes from gravitational contraction and nuclear reactions in approximately equal parts.

T / F 2. In a fission reaction, two small nuclei are combined so that some of the mass is converted into energy.

T / F 3. The Coulomb barrier is the term for the electrical repulsion between two positively charged nuclei.

T / F 4. The Ideal Gas Law states that if volume is held constant, pressure will increase as temperature increases.

T / F 5. The CNO Cycle is a very important means of energy production in low-mass stars like the Sun.

T / F 6. The triple-alpha process can occur only at high temperatures because there are no stable nuclei with eight nucleons.

T / F 7. Degradation is the term used to describe the conversion of a high-energy photon produced in the core to a multitude of low-energy photons at the surface of the Sun.

T / F 8. Neutrinos produced in nuclear reactions make their way to the surface of a star in a random walk in exactly the same way that photons do.

T / F 9. The Sun produces half as many photons as astronomers predict that it should.

T / F 10. Parallax measurements made from ground-based telescopes are just as accurate as those made from space-based telescopes.

T / F 11. If a star has zero proper motion, then it must be moving either directly toward us or directly away from us.

T / F 12. The frequency response of the eye is very similar to the B filter.

T / F 13. The difference between magnitudes in two different filters is called a color index and is useful for learning about the temperature of a star.

T / F 14. The primary reason for the different spectral types is that stars have different compositions.

T / F 15. Since white dwarfs have high surface temperatures and are faint, they must be very small.

T / F 16. In the binary system Sirius A and B, the high-mass star Sirius A is stationary while the low-mass star Sirius B orbits around it.

T / F 17. The inclination of a binary's plane of orbital revolution is typically not known.

T / F 18. The luminosity of a 12 solar mass red giant could easily be calculated with the mass-luminosity relation.

T / F 19. It is often possible to determine the masses of the stars in astrometric binaries.

T / F 20. A spectroscopic binary where the spectral lines of one star are too faint to be seen is known as a single-line spectroscopic binary.

T / F 21. When one star of a spectroscopic binary is moving away from us, its spectral lines are blueshifted.

T / F 22. The inclination angle of the plane of orbit for eclipsing binaries can have any value between 0° and 90°.

T / F 23. Since the two stars in the Algol system are roughly the same size, the deepest eclipse in the light curve occurs when the spectral type B8 star eclipses the spectral type K2 star.

T / F 24. Accretion disks are often powerful sources of X-rays.

T / F 25. A nova involves the accretion of material from a star onto a neutron star where it ignites in a thermonuclear explosion.

T / F 26. The presence of a black hole is the best explanation for a binary system where there is a strong X-ray source and one star is massive but cannot be seen.

T / F 27. Stars in an association likely formed from the same cloud since they appear to be moving as a group but are too far apart to be bound by gravity.

T / F 28. Messier objects are galaxies, nebulae, and clusters that could possibly be mistaken for comets through a small telescope.

T / F 29. The stars in globular clusters must be old because they contain a high percentage of metals that hadn't yet been used up when the galaxy was young.

T / F 30. Harlow Shapley used the distribution of globular clusters to estimate the location of the center of our galaxy.

Unit 12
Star Birth and the Main Sequence

Chapter Objectives

In this chapter we will apply the basic physical principles of gravitational collapse to the nebulae where star birth occurs. The criteria necessary to begin the star formation process will be discussed, and stages in the life of a protostar will be illustrated. The most recent infrared images revealing these previously hidden star birth milestones will be presented. The evolutionary tracks calculated for forming stars will be traced out on the H-R diagram. The limiting masses for stars will be discussed and the brown dwarfs ("stars that missed") will be described.

Progress Checklist

1. Recipe for Stars
- ❑ Molecular Clouds
- ❑ Temperature, Pressure, and Gravity
- ❑ Hydrostatic Equilibrium
- ❑ Jeans Collapse Criterion
- ❑ Fragmentation
- ❑ Sources of Instability

2. Protostars
- ❑ Cocoons
- ❑ Disks and Bipolar Flows
- ❑ T-Tauri Stars
- ❑ Herbig-Haro Objects
- ❑ Motion on the HR Diagram
- ❑ Kelvin-Helmholtz Timescale

3. The Main Sequence
- ❑ Main Sequence Life
- ❑ Width of Main Sequence
- ❑ Dynamical Timescales
- ❑ Limiting Lower Mass
- ❑ Brown Dwarfs
- ❑ Limiting Upper Mass

Keywords

molecular clouds
molecular hydrogen
hydrostatic equilibrium
hydrodynamics
Jeans mass
Jeans density
fragmentation of cloud
spiral density waves
shock waves
protostars
cocoon
EGGs
accretion disks
bipolar flow
T-Tauri stars
Herbig-Haro objects
ejected jets
evolutionary track
Hyashi track
radiative core
Kelvin-Helmholtz timescale
virial theorem
main sequence lifetimes
ZAMS
dynamical timescales
free-fall timescale
expansion timescale
limiting lower mass
brown dwarfs
upper mass limit
Eddington luminosity
Wolf-Rayet stars

Exercise 12-1: Introductory Narrative

Stars are formed in giant clouds of gas and dust. Because it is cold enough in these clouds for two hydrogen atoms to be bound together, they are known as 1) ____________________. The cloud can collapse if it has a mass greater than the 2) ____________________ mass, which is proportional to the temperature and radius and inversely proportional to the average mass of gas particles in the cloud. This collapse may be started by density waves in spiral galaxies or the 3) ____________________ from nearby supernovae. The cloud will typically break up into smaller collapsing clouds in a process known as 4) ____________________. As a cloud collapses, it heats up as a result of the conservation of 5) ____________________. This increase in temperature is accompanied by an increase in the average velocity of the particles that inhibits further collapse. Thus, the rate of collapse is related to the rate of 6) ____________________ to the surface.

It is often difficult to study protostars because they are surrounded by gas and dust, which form a(n) 7) ____________________ around them. In the final stages of collapse, a protostar will be surrounded by an accreting disk of material and have jets of material known as 8) ____________________ streaming out along the axis of rotation.

It is thought that the Sun took about 9) ____________________ years to collapse to the main sequence. The path it traced out moving toward the main sequence on the HR Diagram is known as a(n) 10) ____________________. All main sequence stars are producing energy by fusing 11) ____________________ into helium. The most important quantity in determining a star's life is its mass. Stars on the main sequence have masses ranging from 12) __________ to over 100 solar masses. Both the time of collapse and the time spent on the main sequence decrease as the mass of the star increases.

Exercise 12-2: Luminosities and Lifetimes of Main Sequence Stars*

$$LT = fMc^2$$

$$T = \frac{fMc^2}{L} = \frac{fMc^2}{M^{3.5}}$$

$$T = \frac{fc^2}{M^{2.5}}$$

$$\frac{T}{T_\odot} = \frac{\dfrac{fc^2}{M^{2.5}}}{\dfrac{fc^2}{M_\odot^{2.5}}}$$

$$T = \frac{1}{M^{2.5}}$$

In this exercise we will calculate the luminosity and lifetime of a main sequence star of a given mass. Although we have been using these values for the last two units, here we will see where they actually come from. There are fairly simple formulas for both that make use of solar units, but a calculator will be necessary to complete this exercise.

The luminosity of a star can be calculated using the mass-luminosity relation: $L = M^{3.5}$. We can use this to derive an expression for the lifetime, T, of a star on the main sequence.

We know that the energy a star produces comes from nuclear reactions. Over the course of a star's lifetime, some fraction, f, of its mass gets converted to energy. The process is governed by Einstein's equation $E = mc^2$. Thus, we can write an equation that relates all of the energy that a star produces in its lifetime to the total amount of mass converted to energy.

Rather than estimate the quantity f, let's just assume that it is roughly the same for all stars. So, if we take our expression for stellar lifetime and divide by the same expression for the Sun, the quantity f cancels out. We are left with a simple expression in solar units.

Example: Let's apply these formulas to Sirius, the brightest star in our sky. Sirius has a mass of 2.3 solar masses. The luminosity is given by:

$$L = (M)^{3.5} = (2.3M_\odot)^{3.5} = 18.5L_\odot$$

The lifetime is given by:

$$T = \frac{1}{M^{2.5}} = \frac{1}{(2.3M_\odot)^{2.5}} = 0.12T_\odot$$

The estimated main sequence lifetime of the Sun is 10 billion years, so we can easily convert stellar lifetimes to actual years:

$$0.12T_\odot = 0.12T_\odot\left(\frac{1 \times 10^{10}\ years}{1T_\odot}\right) = 1.2 \times 10^9\ years$$

From this example we can see some general trends in stars. If a star is more massive than the Sun, it will be considerably more luminous. This occurs because the weight of the star pushing down on the core produces much higher core temperatures; thus, nuclear reactions occur at a faster rate. This causes the fuel in the core to be used up more rapidly, however, so more massive stars don't live nearly as long as the Sun.

Directions: Calculate the luminosity (in solar luminosities) and the main sequence lifetimes (in years) for the following stars.

Star #1: Proxima Centauri $M = 0.1\ M_\odot$
Luminosity
Main Sequence Lifetime

Star #2: Rigel $M = 10\ M_\odot$
Luminosity
Main Sequence Lifetime

You can check your answers using the Calculator for the Lifetime on the Main Sequence vs. Mass (**IC 20.1**).

Exercise 12-3: True/False Questions

T / F 1. Since UV radiation easily breaks molecules apart, giant molecular clouds must have considerable amounts of dust in them to shield the molecules from UV radiation from nearby stars.

T / F 2. As a cloud of gas contracts, its temperature will fall as gravitational potential energy is converted into kinetic energy.

T / F 3. As a star contracts, it is in hydrostatic equilibrium.

T / F 4. The Jeans mass increases with the temperature and radius of a cloud of gas.

T / F 5. A large collapsing nebula is likely to fragment into smaller collapsing clouds due to fluctuations that cause subregions of the nebula to exceed the Jeans density.

T / F 6. Shock waves from nearby supernovae may start the collapse of molecular clouds.

T / F 7. Evaporating gaseous globules are dense regions of a nebula where the material has been concentrated due to UV radiation from nearby stars.

T / F 8. Herbig-Haro objects are created when bipolar flows collide with the gas of the interstellar medium.

T / F 9. Collapsing protostars transport energy entirely by radiation.

T / F 10. The Kelvin-Helmholtz timescale states that massive stars take much longer than smaller stars to collapse to the main sequence.

T / F 11. The most important parameter for describing a star's life is its mass.

T / F 12. The width of the main sequence on the HR Diagram is due to differences in stellar masses.

T / F 13. The upper limit for the mass of a star is determined by radiation pressure because it increases much more rapidly with temperature than gas pressure.

T / F 14. The presence of lithium is useful in distinguishing between faint stars and brown dwarfs because lithium is produced in a star's fusion reaction.

T / F 15. Low-mass stars are difficult to find because they are faint and produce most of their radiation in the infrared.

Unit 13
Star Death

Chapter Objectives

What happens when a star runs out of hydrogen fuel in its core and reaches the end of its main sequence lifetime? In this chapter we will explore the stages a star evolves through as it approaches its inevitable death. The three end stages of stars (white dwarf, neutron star, and black hole) will be introduced and the processes leading to their production described. The red giant phase of life for a star like our Sun will be illustrated and the horizontal and asymptotic branches of red giant stars will be described. The types of variable stars that result from the pulsation in size that often accompanies the evolving states of giant and supergiant stars will be delineated. The nuclear processes that result in nucleosynthesis of heavy elements in dying stars will be described.

Progress Checklist

1. End of Main Sequence Life
❏ Lifetime on the Main Sequence
❏ Three End Games
❏ Evolutions away from Sequence
❏ Shell Burning

2. Red Giants
❏ Red Giant Evolution
❏ Red Giant Branch
❏ Thermonuclear Runaways
❏ Helium Flash
❏ Horizontal Branch
❏ Asymptotic Giant Branch

3. Planetary Nebula
❏ Examples
❏ Mass Loss in Red Giants
❏ Ejection on Envolope
❏ Track on HR Diagram

4. Supernovae
❏ Classification
❏ Type I Supernovae
❏ Type II Supernovae
❏ Supernova 1987A
❏ Supernova Candidates
❏ Supernova Remnants

5. Heavy Element Production
❏ Elemental Abundances
❏ Production of Light Elements
❏ Elements up to Silicon
❏ The Iron Peak
❏ The s-Process
❏ The r-Process

6. White Dwarfs
❏ White Dwarfs
❏ Size and Density
❏ Chandrasekhar Mass
❏ Novae

7. Variable Stars
❏ Variable Stars
❏ Cepheid Variables
❏ RR Lyrae Variables
❏ Pulsation Timescales
❏ Long-Period Variables
❏ Instability Strip

Keywords

main sequence lifetime
red giant
white dwarf
planetary nebula
horizontal branch
asymptotic giant branch
triple alpha process
electron degeneracy
fermions
bosons
exclusion principle
thermonuclear runaway
helium flash
helium burning shell
superwind
Chandrasekhar mass limit
nova
Cepheid variables
RR-Lyrae variables
pulsation timescales
hydrodynamical response times
long period red variables
instability strip
supergiant
supernova, type I, type II
gravitational core collapse
shock wave
neutron star
black hole
nuclear density
progenitor star
supernova remnants
nucleosynthesis
iron peak
radiative capture
s-process
r-process
neutron capture
beta decay
chart of the nuclides
beta stability valley

Exercise 13-1: Introductory Narrative

Our Sun will live for about 1) ____________________ years on the main sequence burning hydrogen into helium. When the Sun has exhausted the hydrogen in its core, it will move toward the 2) ____________________ of the HR Diagram. Its energy will now come from 3) ____________________ burning in a shell surrounding the helium ash core, and since the energy source is now closer to the surface, the Sun will expand and become a(n) 4) ____________________. The helium core has no energy source to support the weight of the upper layers of the Sun, so it will contract and its temperature will rise. When the temperature of the now degenerate core reaches about 100,000,000 K, the triple-alpha process will vigorously begin in what is known as the 5) ____________________. Since the primary energy source is now back in the core, the Sun will contract and move back toward the main sequence. The Sun will now stably burn helium in its core in a location of the HR Diagram known as the 6) ____________________. The Sun will burn helium in its core (~7% of its life) for a much shorter time than it will burn hydrogen (~93% of its life). When the Sun runs out of helium in its core, it will again become a(n) 7) ____________________ as it burns helium in a shell surrounding the core of carbon ash. The Sun has insufficient mass to ever be hot enough to fuse carbon. The Sun will now become so large that it will blow off most of its outer layers into space in a phenomenon known as a(n) 8) ____________________. This exposes the hot inner layers of the Sun, causing it to move immediately to the far 9) ____________________ of the HR Diagram. The Sun cannot get any more energy from fusion and greatly resembles the dying embers of a fire. It will contract due to its own weight until gravity is halted by 10) ____________________. Then the Sun will slowly cool down and become a(n) 11) ____________________ about the size of the 12) ____________________.

Exercise 13-2: Using Pulsating Variable Stars as Distance Indicators*

Many of the later stages in a star's life can be very useful as distance indicators. All of the distance indicators we will use here are called "standard candles," which means that we have a good idea of how intrinsically bright they are—we know their absolute magnitude. We can observe their apparent magnitude and compare the two to get distance.

RR Lyrae stars are low-mass pulsating variables. Although the absolute magnitude of RR Lyrae stars varies slightly with pulsation period, one can see from the accompanying diagram that the variation is very small. For our purposes it is sufficient to assume that all RR Lyrae stars have absolute magnitudes of $M = +0.5$.

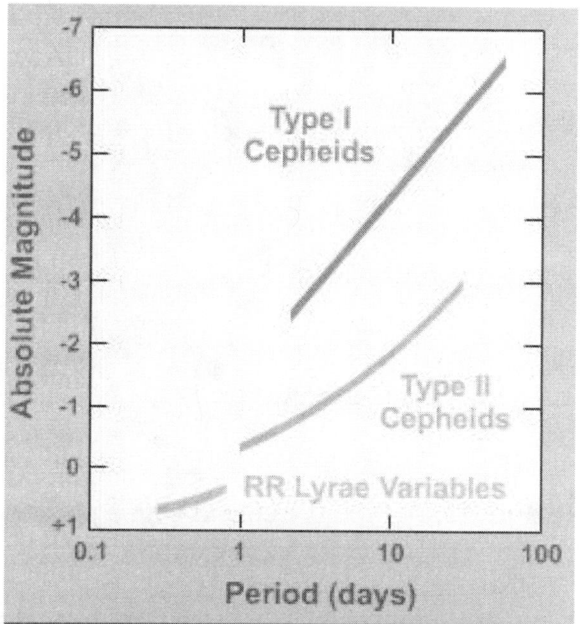

Period-Luminosity Relations

For the more massive Cepheid variables, we will need to observe the pulsation period of the star and then use it to obtain the absolute magnitude from the chart. It is also necessary to know the metallicity of the star since there is a considerable difference in the period-luminosity relation for Type I and Type II Cepheids.

Example I: A Type I Cepheid variable is observed to have an apparent magnitude of $m = 6$ and a pulsation period of 50 days. How far away is the star?

Using the period-luminosity relation, we can see that a Type I Cepheid with a period of 50 days has an absolute magnitude of approximately –5. Thus, the distance modulus $m - M = 6 - (-5) = +11$. By extrapolating the pattern in the accompanying table, we can conclude that a distance modulus of 11 corresponds to a distance of 1600 parsecs (pc).

Distance Modulus ($m - M$)	Distance (d in pc)
0	10
1	16
2	25
3	40
4	63
5	100
6	160
7	250
8	400
9	630
10	10^3
15	10^4
20	10^5

Example II: An RR Lyrae star has an apparent magnitude of 8. How far away is it?

$$m - M = 8 - (0.5) = +7.5$$

We can interpolate between the distance modulus values for 7 and 8 and get a distance of approximately 325 pc.

Directions: Estimate the distance to each of the following stars.

Star #1: An RR Lyrae star with $m = 9.5$
Distance

Star #2: A Type II Cepheid with a pulsation period of 5 days and $m = 4$
Distance

Star #3: A Type I Cepheid with a pulsation period of 10 days and $m = 6$
Distance

Exercise 13-3: True/False Questions

T / F 1. Stars leave the main sequence when approximately 10% of their original supply of hydrogen has been used up.

T / F 2. All of the red dwarfs that have ever formed are still on the main sequence because the universe is not yet old enough for them to have evolved off of it.

T / F 3. When the Sun runs out of hydrogen in its core, it will start burning helium.

T / F 4. The horizontal branch of the main sequence is a relatively stable region where core helium burning takes place.

T / F 5. Degenerate gases have the important property that their pressure is independent of temperature.

T / F 6. Planetary nebulae occur when the planets surrounding evolving stars are destroyed and the debris forms a shell around the star.

T / F 7. White dwarfs typically have hot surface temperatures due to the fusion of helium into carbon.

T / F 8. Chandrasekhar's limit states that all white dwarfs have to be more massive than 1.4 solar masses.

T / F 9. RR Lyrae stars are useful as distance indicators because they all have approximately the same intrinsic brightness.

T / F 10. Cepheid variables are useful as distance indicators because their luminosity is larger for stars with shorter pulsation periods.

T / F 11. The spectra from Type Ia supernovae do not show hydrogen lines because no hydrogen is present in the iron core.

T / F 12. The only difference between a Type Ia supernova scenario and a nova scenario is the rate at which matter accretes onto the surface of the white dwarf.

T / F 13. Astronomers were surprised to learn that the progenitor of Supernova 1987A was a blue supergiant instead of a red supergiant.

T / F 14. The nuclei with the highest nuclear binding energies are the light nuclei produced in the big bang.

T / F 15. The building of nuclei through the rapid capture of neutrons known as the r-process is thought to occur in Type Ia supernovae.

Conceptual Map 4
Stellar Evolution

In this exercise we will study the evolutionary tracks of stars on the Hertzsprung-Russell Diagram. In the following diagram, tracks are shown from the main sequence for stars of approximately 0.1, 1, and 20 solar masses. The evolutionary tracks for stars of a certain initial mass vary due to differences in composition and the amount of mass loss at various stages of their lives. Thus, these tracks only represent one possible evolution of the star.

First you should identify which track corresponds to which stellar mass. The terms below all correspond to a particular stage in one of the three stars' lives. Some of the terms apply to more than one star. Place the letter for each term at the appropriate location(s) on the appropriate evolutionary track(s).

A. Main Sequence Star
B. Helium Flash
C. Carbon Detonation
D. Red Giant
E. Horizontal Branch
F. White Dwarf
G. Planetary Nebula
H. Supernova
I. Fusion of Heavy Elements
J. Our Sun Today

If one could survey all of the many stars in the Milky Way, do you think we could find a star at virtually every point on the three evolutionary tracks? For which sections of which tracks would it be hard to find a star? Is there a part of one of the tracks for which no stars could be found in the Milky Way? Explain.

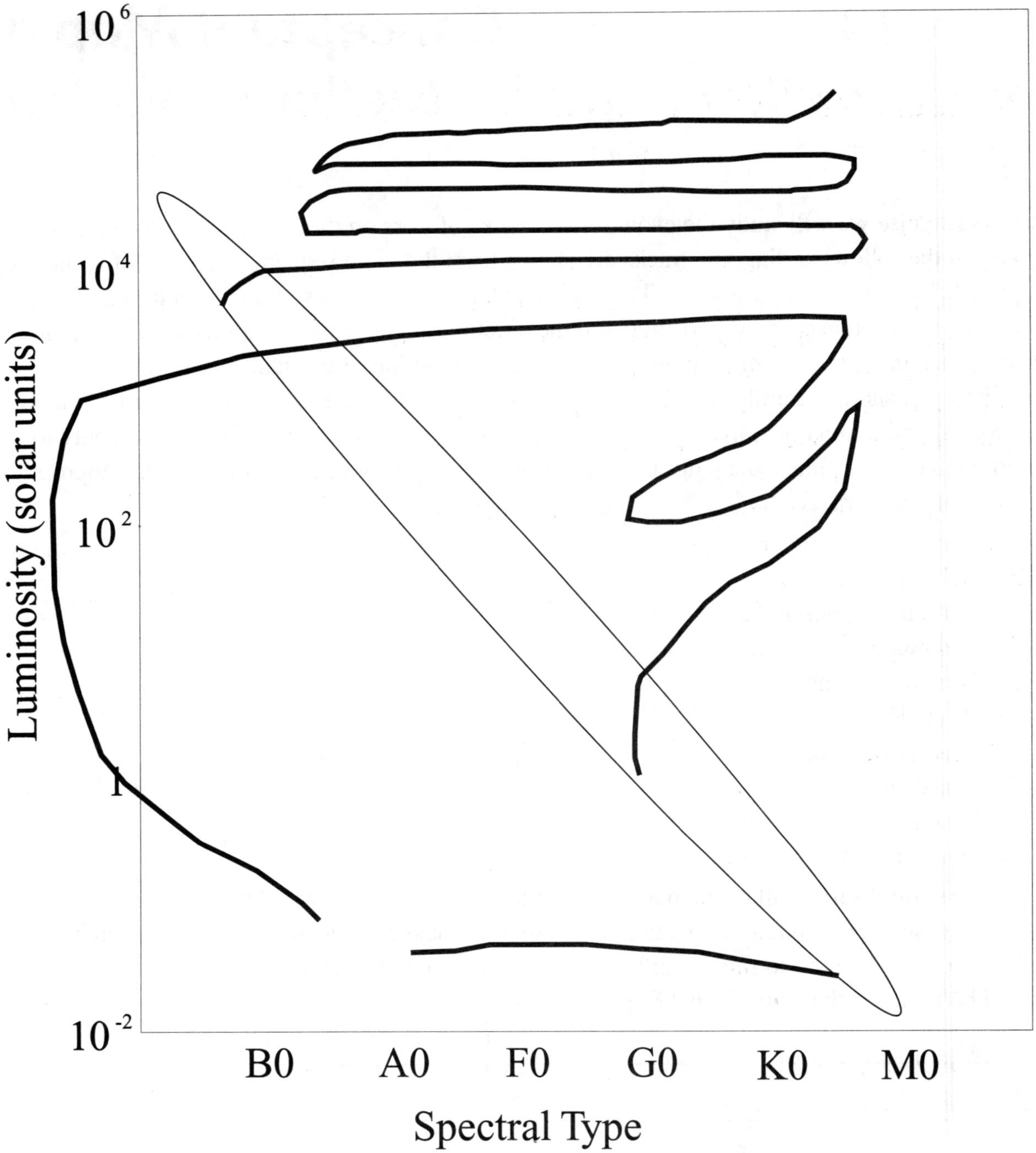
10^6
10^4
10^2
1
10^-2
Luminosity (solar units)
B0
A0
F0
G0
K0
M0
Spectral Type

Unit 14
Neutron Stars and Black Holes

Chapter Objectives

At the end of their lives, the most massive stars will have collapsing cores that will exceed the Chandrasekhar limit and therefore these stars cannot leave white dwarfs as their stellar corpses. This chapter will examine the two possible core remnants that result when massive stars die: neutron stars and (for the most massive stars) black holes. The neutronization process that occurs during the formation of a neutron star will be explained and the basic structure of the neutron star that results will be described. The discovery of pulsars and the subsequent identification of these rapidly pulsing objects as rotating neutron stars will be chronicled. The upper limit to the mass of a neutron star will be shown to be about three solar masses; thus, a very massive star will undergo a core collapse that cannot be stopped, resulting in a black hole. The simple characteristics of a black hole will be described and the remarkable effects on the surrounding space-time will be illustrated. The search for stellar black holes and the list of the best candidates for black holes in binary star systems will be discussed.

Progress Checklist

1. Neutron Stars
- ❑ Neutron Stars
- ❑ Size and Density
- ❑ Production of Neutron Stars
- ❑ Surface Gravity
- ❑ Magnetic Field
- ❑ X-Ray Bursts

2. Pulsars
- ❑ Discovery
- ❑ The Lighthouse Model
- ❑ Rotation Rates
- ❑ The Crab Pulsar
- ❑ Binary Pulsar
- ❑ Magnetars

3. Stellar Black Holes
- ❑ The Event Horizon
- ❑ The Singularity
- ❑ Spacetime Paths
- ❑ Properties of a Black Hole
- ❑ Black Hole Candidates
- ❑ Rotation Black Holes

Keywords

neutron stars
Chandrasekhar limit
electron capture
neutronization
pulsars
lighthouse model
Crab pulsar
synchrotron acceleration
starquakes
millisecond pulsars
binary pulsar
general relativity
curved spacetime
gravitational waves
magnetars
soft gamma ray repeaters
X-ray burster
event horizon
gravitational red shift
Schwarzschild radius
singularity
naked singularity
photon sphere
Einstein ring
ergosphere
Kerr solution
no hair theorem
frame dragging
Cygnus X-1

Exercise 14-1: Introductory Narrative

When low-mass stars like the Sun end their lives, they become 1) ____________________, but more massive stars become either neutron stars or black holes. Neutron stars are formed from stellar remnants more massive than 2) ____________________, up to possibly 3 solar masses although this upper limit is not well known. At these larger masses, electron pressure is insufficient to support the weight of the material, so electrons and 3) ____________________ are forced together to form neutrons and the object collapses to about the size of 4) ____________.

Neutron stars have very powerful 5) ____________________ fields. Due to the conservation of angular momentum, neutron stars also rotate 6) ____________________. The combination of these two effects causes strong electric fields that accelerate electrons away from the surface near the magnetic poles, producing 7) ____________________ radiation in a pair of beams. Because the magnetic axis doesn't coincide with the rotation axis, these beams trace out a corkscrew pattern as the neutron star rotates. If one of the beams happens to cross our line of sight, we perceive the neutron star to be a(n) 8) ____________________. This explanation for the phenomenon is known as the 9) ____________________ model.

More massive stellar remnants become black holes. The matter collapses down to an object of zero size known as a(n) 10) ____________________. Gravity is so strong near a black hole that the escape velocity is equal to the speed of light. The region surrounding a black hole from which light cannot escape is known as the 11) ____________________. Black holes are extremely simple because all information concerning their constituent matter is destroyed as they form. Their only distinguishing characteristics are mass, 12) ____________________, and angular momentum.

Since light cannot escape from black holes, the only way to detect them is indirectly through the effects they have on surrounding matter. This typically involves detecting the 13) ____________________ that are produced from friction as material spirals into a black hole.

Exercise 14-2: True/False Questions

T / F 1. An electron and a proton can combine to form a neutron and an electron neutrino.

T / F 2. Neutron stars can have masses between 0.8 and about 3 solar masses.

T / F 3. Neutron stars are thought to be the cores of imploded Type I supernovae.

T / F 4. Supernovae explosions are thought to be asymmetric due to the high space velocities of neutron stars.

T / F 5. Neutron stars have extremely high surface gravity and strong magnetic fields.

T / F 6. When matter accretes onto the surface of a neutron star from another star in a binary system, a nova may occur.

T / F 7. When first detected, the regularity of the signals from pulsars led some to suggest they were "little green men."

T / F 8. The magnetic axis of a neutron star is well aligned with its rotation axis.

T / F 9. As a pulsar gets older, its rate of spinning gradually increases as further gravitational collapse occurs.

T / F 10. The sudden increases in pulsar periods known as glitches are probably due to seismic events.

T / F 11. Millisecond pulsars spin extremely rapidly because they are very young and haven't had time to lose much energy.

T / F 12. The region surrounding a black hole for which the escape velocity is the speed of light is known as the event horizon.

T / F 13. Light can travel in a circular orbit around a black hole.

T / F 14. The statement "black holes have no hair" refers to the smoothness of the photon sphere.

T / F 15. Black holes are most often identified by X-rays produced by material spiraling into them.

Unit 15
The Milky Way

Chapter Objectives

The Universe is composed of billions of separate collections of stars known as galaxies, and our home galaxy is an example of the spiral type of galaxy. Our Sun is located in the outer reaches of the Milky Way Galaxy and is one of the several hundred billion stars that comprise this typical spiral. In this chapter we will study the parts of the Milky Way Galaxy: the halo, the disk, and the central bulge. The views of our galaxy in different wavelengths will be illustrated and the resulting evolution of our perception of our home galaxy will be discussed. The spiral structure found in the disk portion of our galaxy will be described and the mechanisms that may have caused this spiral pattern will be explored. The concept of dark matter and its role in the dynamics of our galaxy will be introduced.

Progress Checklist

1. Observing the Galaxy
- ❑ The Milky Way
- ❑ Our Evolving Perception
- ❑ Measuring the Galaxy
- ❑ Visible Light View
- ❑ IR View
- ❑ Radio Frequency View

2. Components of the Galaxy
- ❑ Components
- ❑ Galactic Disk
- ❑ Visible Halo
- ❑ Galactic Bulge
- ❑ Magnetic Field and Cosmic Rays
- ❑ Dark Matter Halo

3. Rotation and Spiral Structure
- ❑ Spiral Structure
- ❑ Density Waves
- ❑ Self-Sustaining Star Formation
- ❑ Rotation Curves and Mass

4. The Interstellar Medium
- ❑ The Interstellar Medium
- ❑ Interstellar Gas
- ❑ HI and HII Regions
- ❑ Interstellar Dust
- ❑ Nebulae
- ❑ Aluminum-26

Keywords

Milky Way
Milky Way galaxy
grindstone model
Cepheid variables
spiral galaxy
kiloparsec
megaparsec
galactic coordinate system
galactic longitude
galactic latitude
galactic poles
infrared telescope
radio telescope
neutral hydrogen
21 cm line
spin-flip transition
central bulge
halo
disk
spiral arms
spherical component
galactic magnetic field
dark matter
Population II
Population I
metal poor
metal rich

interstellar medium
Sagittarius A*
polarization of light
cosmic rays
H-II regions
spiral tracers
density waves
self-sustaining star formation
differential rotation
grand design spirals
rotation curve
emission nebulae
reflection nebulae
carbon monoxide
coronal gas of ISM
HI region
interstellar dust grains
dark (absorption) nebulae
interstellar extinction
interstellar reddening
forbidden transitions
metastable state
Aluminum-26

Exercise 15-1: Introductory Narrative

Our Solar System is one of approximately 1) ____________________ stars that along with gas and dust make up the Milky Way Galaxy. Since we are located inside the Milky Way, it was difficult for astronomers to discern the nature of the galaxy. Two major breakthroughs occurred in the 1920s using Cepheid variable stars as distance indicators. 2) ____________________ determined the scale of the Milky Way. He mapped the location of globular clusters, and by assuming that they were distributed in a spherically symmetric pattern, he calculated the distance to the center of our galaxy. 3) ____________________ found that the distance to the Andromeda Galaxy is much larger than the size of our own galaxy. He showed that galaxies were really "island universes," huge isolated collections of stars.

The Milky Way has three visible components known as the disk, the halo, and the 4) ____________________. These components are a result of the way our galaxy formed from a large cloud of gas and dust. Originally, the cloud was spinning rather 5) ____________________ and had a(n) 6) ____________________ abundance of metals. The first stars to form were population 7) ____________________, spherically distributed, and in randomly oriented elliptical orbits. As the galaxy contracted due to gravity, it began to spin more 8) ____________________ and formed a disk where most of the gas and dust became concentrated. More recent stars that have formed there are population 9) ____________________ and travel in circular orbits in the plane of the disk of our galaxy.

We can learn about the distribution of matter in our galaxy by looking at the orbits of stars. The graph of the orbital velocities of stars versus their distance from the galactic center is called a(n) 10) ____________________. It shows that orbital velocities increase toward the outskirts of the Milky Way. For these stars to be orbiting so rapidly, their orbits must enclose a large amount of mass that astronomers don't see. Thus, most of the matter of our galaxy is unseen and is known as 11) ____________________.

Exercise 15-2: Components of the Milky Way Galaxy

Directions: Below is a list of facts and observations for 15 different stars. Indicate whether the data suggest that the star belongs to the disk (D) or the halo (H) of the Milky Way

______ 1. This star contains 3% metals.

______ 2. This star contains 0.4% metals.

______ 3. This star is in the open cluster known as the Pleiades.

______ 4. Doppler shifts indicate that this star has a high radial velocity.

______ 5. Parallax is large for this star, yet the proper motion is very small.

______ 6. This star is part of an association in the Orion spiral arm.

______ 7. This star is an RR Lyrae star in the globular cluster M3.

______ 8. This star has a very elliptical orbit around the galaxy.

______ 9. This star is a Type II Cepheid variable.

______ 10. This star is an O3 star

______ 11. This star formed when our galaxy had a spherical shape.

______ 12. This star is far away and has an high galactic latitude.

______ 13. This star was formed by "self-sustaining star formation."

______ 14. This star is between a giant molecular cloud and an HII region.

______ 15. The spectra of this star are weak in the lines of heavier atoms.

Exercise 15-3: True/False Questions

T / F 1. For observers in the northern hemisphere, the summer sky's Milky Way is brighter than the winter sky's Milky Way because it is in the direction of the center of the galaxy.

T / F 2. Galactic longitude is the angle between the plane of the disk of our galaxy and a line from our Solar System to the object of interest.

T / F 3. We can view the center of our galaxy better in infrared light than visual because the hot stars there produce more light in this wavelength band.

T / F 4. A population I star is likely to be much older than a population II star.

T / F 5. The halo of our galaxy has a diameter at least as large as the disk's diameter.

T / F 6. Halo stars have circular coplanar orbits.

T / F 7. Because elongated dust particles spin so that they are aligned with any magnetic fields that are present, scattered light is polarized and can be used to study the magnetic field.

T / F 8. The spiral arms of our galaxy are bright because of the supernovae that occur in them.

T / F 9. The definitive characteristic of a spiral tracer is that it must be short-lived so that it cannot move more than the width of a spiral arm in its lifetime.

T / F 10. It is difficult to explain flocculent spiral galaxies using the density wave theory for spiral arms.

T / F 11. The fact that rotational velocities increase at the edge of the Milky Way proves that many black holes are present there.

T / F 12. Although the interstellar medium typically has very low density, it makes up about 15% of the total mass of visible matter in the Milky Way.

T / F 13. Some clouds of gas are known as HII regions because two hydrogen atoms are loosely bound together.

T / F 14. Interstellar reddening occurs because the gas of the interstellar medium preferentially scatters blue light more than red light.

T / F 15. Forbidden transitions occur in nebulae because the density is so low that collisions between atoms are very infrequent.

Unit 16
The Galaxies

Chapter Objectives

The visible Universe contains billions of galaxies that collect together in clusters, superclusters, bubbles, and "great walls" that are the large scale structure of our Universe. In this chapter we will describe the three main classes of galaxies: spiral, elliptical, and irregular galaxies. The differences between the galaxy types will be illustrated and the suggested reasons behind these differences will be explored. The clusters and superclusters of galaxies will be described and the effects of galaxy interactions and collisions in the crowded clusters of galaxies will be shown. The large scale soap bubbles, voids, and walls that appear in three dimensional structural "maps" of our Universe will be introduced.

Progress Checklist

1. Hubble Classification
❑ The Tuning Fork Diagram
❑ Elliptical Galaxies
❑ Spiral Galaxies
❑ Irregular Galaxies
❑ Summary of Galaxy Properties
❑ Galactic Evolution?

2. Clusters and Superclusters
❑ Groups of Galaxies
❑ Clusters of Galaxies
❑ Masses of Clusters
❑ Dark Matter
❑ Superclusters
❑ The Great Attractor

3. The Expanding Universe
❑ Cosmic Distance Ladder
❑ Red Shifts
❑ Hubble Law
❑ Age of the Universe
❑ Look-Back Times
❑ The Most Distance Objects

4. Soap Bubbles and Voids
❑ Redshift Surveys: The 3D Universe
❑ Soap Bubbles and Voids
❑ Great Walls
❑ Summary: 3D Structure
❑ Computer Simulations
❑ Summary of Distance Scales

5. Interacting Galaxies
❑ Evidence for Galaxy Collisions
❑ Starburst Galaxies
❑ Cannibalism and Mergers
❑ Interactions and Evolution
❑ The Early Universe
❑ Andromeda Is Coming

Keywords

Hubble Classification
tuning fork diagram
elliptical galaxies
spiral galaxies
barred spirals
irregular galaxies
giant elliptical
dwarf elliptical
peculiar galaxies
galactic cannibalism
clusters of galaxies
superclusters
groups of galaxies
The Local Group
Andromeda Galaxy
Large Magellanic Cloud
Small Magellanic Cloud
Sagittarius Dwarf

rich clusters
poor clusters
Virgo cluster
Coma cluster
dark matter
X-ray gas
baryonic matter
non-baryonic matter
supersymmetric particles
hot dark matter
cold dark matter
relativistic matter
non-relativistic matter
The Local Supercluster
Hubble flow
peculiar velocity
The Great Attractor
standard candle
Tulley-Fisher Relation
type Ia supernova
Olber's Paradox
Cepheid variables
galactic red shift
red shift parameter
Hubble's Law
Hubble constant
Hubble time
look-back times
pencil surveys
voids
Great Wall
Southern Wall
colliding galaxies
starburst galaxies
Cartwheel Galaxy

Exercise 16-1: Introductory Narrative

Hubble classified galaxies morphologically into groups. He called two of the groups spiral and elliptical; those that didn't fit into either of the previous groups were known as 1) ______________________________ galaxies. The spiral galaxies can be further divided into normal spirals and those that have 2) ____________________ running through their centers. Both the spiral and the elliptical groups are then further subdivided by their shape. Elliptical galaxies are classified as E0 for those that are 3) ____________________ to E7 for highly flattened galaxies. Spirals are further classified as Sa (or SBa) for galaxies with 4) ________________ nuclei and tightly wound spiral arms to Sc (or SBc) for galaxies with small nuclei and 5) ____________________ wound spiral arms. All of these types are organized into a structure known as the 6) ____________________.

Types of galaxies differ in other ways besides shape. Spiral galaxies have considerably more gas and 7) ____________________ than do elliptical galaxies. Since these are the materials needed for 8) ____________________, we see massive young stars that have formed very recently in spirals but not in ellipticals. Consequently, spirals are considerably brighter on average than elliptical galaxies. Thus, the majority of galaxies that we see are spirals even though ellipticals are really the most abundant type of galaxy. Another difference is that ellipticals are more commonly found in regions of space that are densely packed with galaxies. This suggests that galaxies 9) ____________________ due to interactions with their neighbors.

Galaxies are assembled on large scales in groups, clusters, and 10) ________________. All of these are held together by the mutual 11) ____________________ attraction between the galaxies. However, only about 1/10 of the masses needed to keep the clusters of galaxies held together is emitting visible light, which indicates the presence of 12) ______________.

Exercise 16-2: Distances to Galaxies

In this exercise we will use supernovae as standard candles to estimate the distance to remote galaxies. The ideas being used here are very similar to those in Exercise 21-2 where we used pulsating variable stars as distance indicators. Astronomers have a good idea what the absolute magnitude of a supernova is at its peak brightness. They can then compare this peak absolute magnitude with the observed peak apparent magnitude and calculate the distance. Because supernovae are extremely luminous, this technique can be used on galaxies over 1000 Mpc away, whereas Cepheid variables are limited to about 20 Mpc.

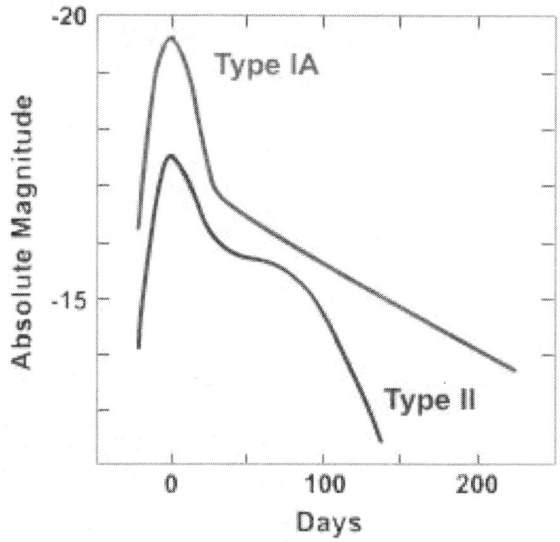

Typical Supernova Light Curves

Example: A Type II supernova is observed for several weeks, and its light curve has a peak apparent magnitude of 13.0. We know from the graph shown to the right that Type II supernovae peak with absolute magnitudes of approximately 17.5.

$$m - M = 13.0 - (-17.5) = 30.5$$

Using the chart to the right, one can see that 30.5 would be between 1.0×10^7 and 1.6×10^7. So let's call our final answer 1.4×10^7 or 14 Mpc.

Galaxy #1: A Type II supernova is observed with a peak apparent magnitude of 20.5. Estimate the distance to this galaxy.

Distance =

Galaxy #2: A Type IA supernova is observed with a peak apparent magnitude of 14.3. Estimate the distance to this galaxy.

Distance =

Distance Modulus $(m - M)$	Distance (d in pc)
0	10
1	16
2	25
3	40
4	63
5	100
6	160
7	250
8	400
9	630
10	10^3
15	10^4
20	10^5
25	10^6
30	10^7
35	10^8
40	10^9

Exercise 16-3: Galaxy Classification

In this exercise we will look at actual pictures of galaxies and try to classify them according to the Hubble Tuning Fork Diagram. This classification breaks galaxies up into the large groups called spirals (which has 6 subclasses) and ellipticals (which has eight subclasses). There are two additional classifications: S0, which serves as a transition between the spiral and elliptical groups, and IRR (irregular galaxies), which is a catchall for galaxies that don't fit into any of the other groups.

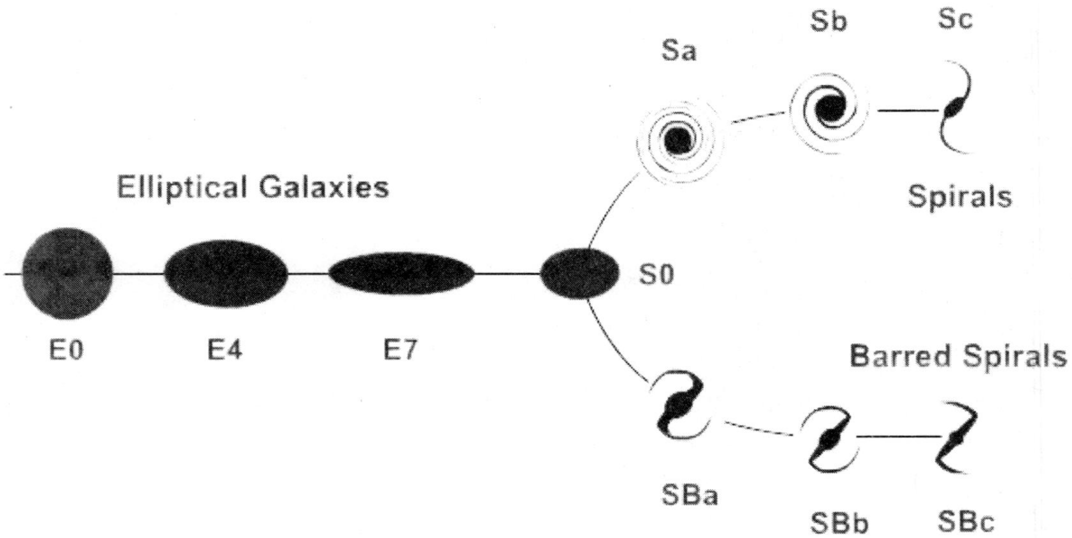

Hubble Classification of the Galaxies

The following flowchart illustrates the thought processes that should be used to accomplish this procedure. The diamond symbols indicate a "fork in the road" where a decision must be made concerning classification. One should first ask, "Is there a disk component?" If a disk is present, the galaxy is a spiral and must be further categorized based on the winding of the spiral arms, nucleus size, and whether a bar exists in the center of the galaxy. If no disk is present, one must next ask, "Is it spherical or elliptical in shape?" If the answer is yes, the galaxy is an elliptical and must be further classified based upon its exact shape. If no, then the galaxy must be an irregular.

Galaxy classification is not an exact procedure. There will always be some imprecision when deciding the exact classification. In addition, because galaxies are viewed in random orientations, it is occasionally difficult to determine even the basic group to which the galaxy belongs. For example, spiral galaxies that are viewed from a perspective nearly in the plane of the disk are particularly troublesome. Since one cannot see the spiral arms, classification relies more heavily on the size of the nucleus and whether gas and dust are visible.

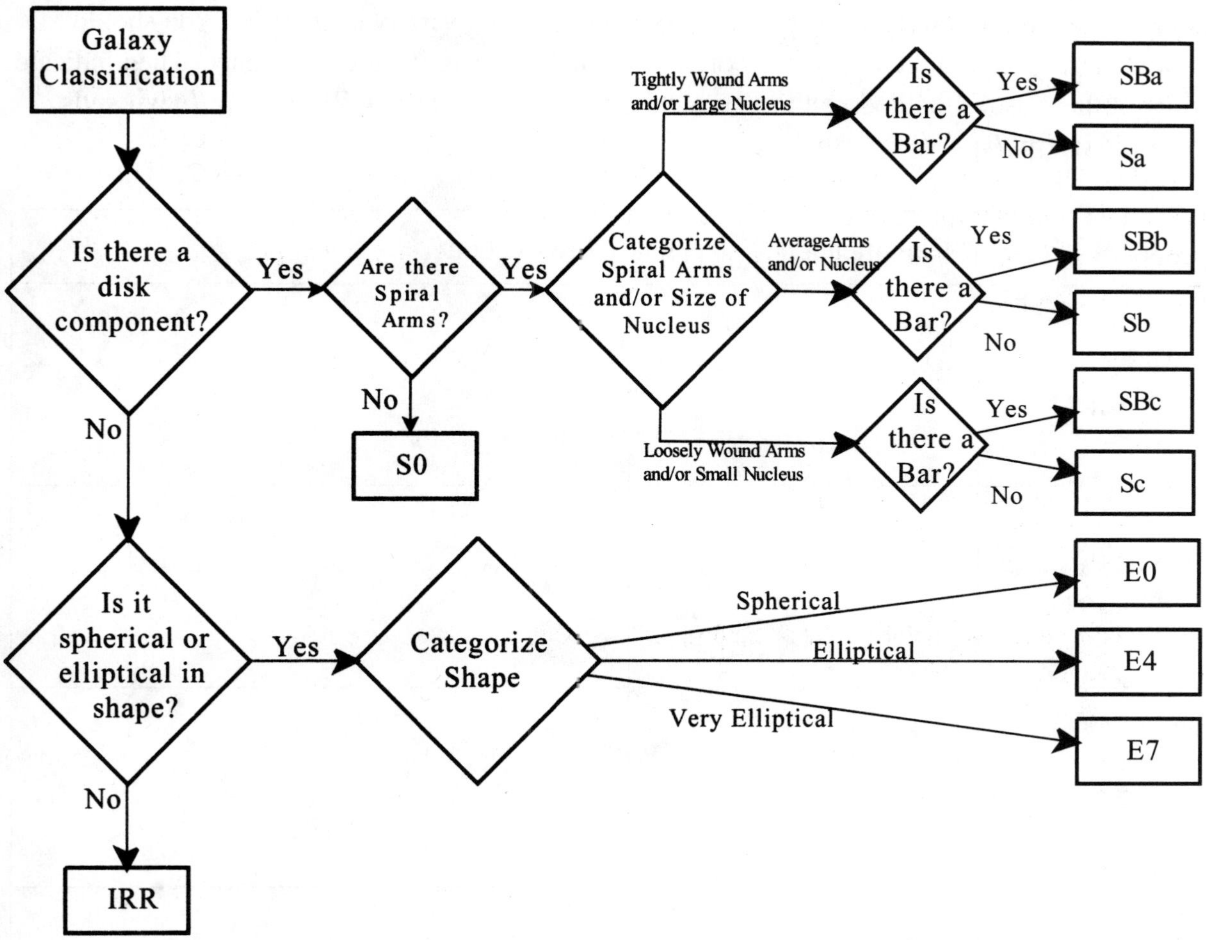
Galaxy Classification
Is there a disk component?
Yes
No
Are there Spiral Arms?
Yes
No
S0
Categorize Spiral Arms and/or Size of Nucleus
Tightly Wound Arms and/or Large Nucleus
AverageArms and/or Nucleus
Loosely Wound Arms and/or Small Nucleus
Is there a Bar?
Yes
No
SBa
Sa
SBb
Sb
SBc
Sc
Is it spherical or elliptical in shape?
Yes
No
IRR
Categorize Shape
Spherical
Elliptical
Very Elliptical
E0
E4
E7

Directions: The following table givs URLs for images of a variety of galaxies. You should look at each galaxy and classify it according to Hubble's classification sequence. Then indicate some of the reasons that led you to your decision. The base of the URL is **http://www.seds.org/messier/jpg/** and is the same for each image.

URL	Classification	Justification
Base+ **m64.jpg**		
Base+ **m82.jpg**		
Base+ **m87.jpg**		
Base+ **m33.jpg**		
Base+ **m32.jpg**		
Base+ **m91.jpg**		

The known classifications and other information about these galaxies can also be found on the seds site. For example, for the first galaxy the URL would be **http://www.seds.org/messier/m/m064.html**

Exercise 16-4: True/False Questions

T / F 1. The Hubble Tuning Fork Diagram is useful because it shows how galaxies evolve from E0 to either Sc or SBc.

T / F 2. Elliptical galaxies have a far greater range of sizes than do spiral galaxies.

T / F 3. Elliptical galaxies typically have far more gas and dust than do spirals.

T / F 4. The spiral arms of a spiral galaxy are visible due to the large concentration of small faint stars found there.

T / F 5. Approximately 70% of the galaxies we see are spiral, so they must be the most common type of galaxy.

T / F 6. Since the Andromeda Galaxy is about 3 million light-years away from us, the most recent information we can obtain about it is 3 million years old.

T / F 7. It is difficult to determine the number of members in our Local Group of galaxies because the dust in the plane of the Milky Way may obscure some of them.

T / F 8. If a typical cluster of galaxies has a mass to light ratio of about 400, the cluster must contain many bright galaxies.

T / F 9. Because there are so many neutrinos in the Universe, if only one of the families of neutrinos had a nonzero mass, that could possibly be sufficient to answer the dark matter question.

T / F 10. An extremely massive object known as the Great Attractor causes a large-scale streaming of galaxies in the direction of Centaurus known as the Hubble flow.

T / F 11. Olber's Paradox or "Why is the night sky dark?" is resolved by the expansion of the universe.

T / F 12. One can estimate the age of the universe by calculating the reciprocal of the Hubble constant.

T / F 13. Redshift surveys of galaxies have shown structure on a scale even larger than that of superclusters.

T / F 14. In starburst galaxies the production rate of new stars may be as high as two to three per year.

Conceptual Map 5
The Cosmic Distance Ladder

In this assignment we will complete a logarithmic distance axis from one parsec to 1000 Mpc and describe the methods of distance determination that are used at various distances. This exercise will require you to tie together concepts from the last few chapters.

The general format of this Conceptual Map is a box with a small area on top to contain the name of the distance determination technique and a larger area below to describe exactly how the method is used. In each of these boxes either the name of the technique or the description are already completed for you and you are expected to add the other component.

Distance Determination Technique
Description of Technique

Our distance axis covers the Milky Way galaxy and the Universe. One could have extended the lower range of the distance axis into the solar system and discussed the determination of distances to the planets. An example box describing radar is shown to the right.

RADAR
In this technique, radio waves are transmitted toward a planet or asteroid. The time between transmission and reception of the returning echo is measured. The distance to the body (times 2) is then found by multiplying the elapsed time by the speed of light.

A line is drawn connecting the box to an appropriate distance on the axis where the technique might be applied. Since all commonly used techniques are valid at a range of distances, we will associate a technique with one of the larger distances at which it is typically used. Note that the ranges of the various techniques often overlap. For example, there are RR Lyrae stars which are close enough so that parallax measurements may be made for them. Thus, the distances to the nearer RR Lyraes are "calibrated" using parallax measurements. Distances can then be determined to RR Lyraes that are too far away for parallax. This is where the term "cosmic distance ladder" originates. Each method representing a rung of the ladder is used to refine the technique applied at greater distances which represents the next higher rung of the ladder.

After you have filled in the boxes describing the distance determination methods, you should add some reference objects from the Milky Way and the Universe to the distance axis. Add labels for the following objects at the appropriate locations on the distance axis.

- Alpha Centauri
- Center of the Milky Way
- Andromeda Galaxy
- Virgo Supercluster

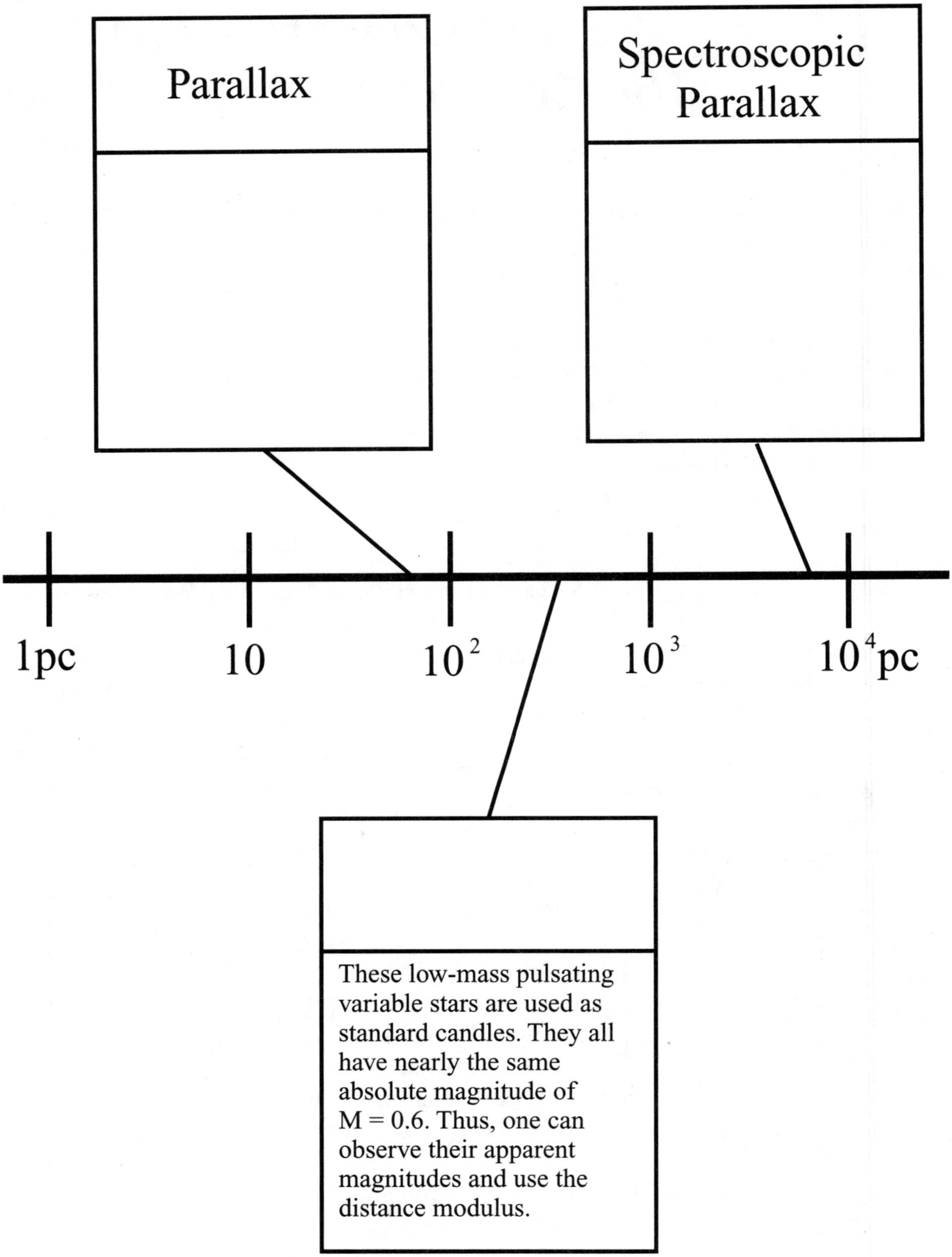
Parallax
Spectroscopic Parallax
1pc
10
10²
10³
10⁴pc
These low-mass pulsating variable stars are used as standard candles. They all have nearly the same absolute magnitude of M = 0.6. Thus, one can observe their apparent magnitudes and use the distance modulus.

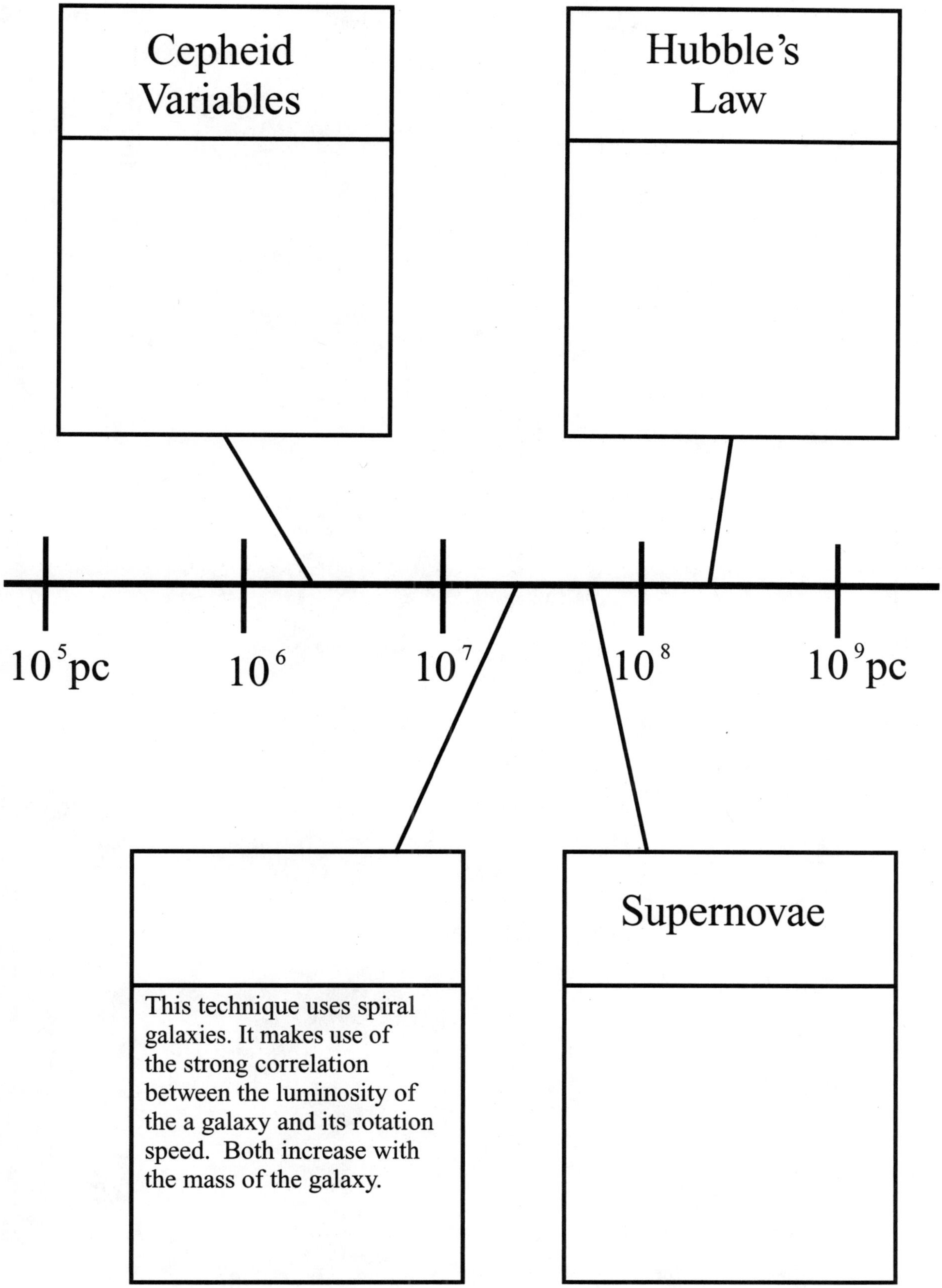
Cepheid Variables
Hubble's Law
10^5pc
10^6
10^7
10^8
10^9pc
This technique uses spiral galaxies. It makes use of the strong correlation between the luminosity of the a galaxy and its rotation speed. Both increase with the mass of the galaxy.
Supernovae

Unit 17
Active Galaxies and Quasars

Chapter Objectives

Some galaxies have very active, energetic, small nuclei that emit hundreds to thousands of times more energy than an entire normal galaxy emits. The historical development of our knowledge of quasars and related galaxies will be delineated in this chapter. The initial controversy over the distance to and thus the nature of these quasi-stellar objects and the evidence pointing to the cosmological interpretation will be presented. The current consensus concerning the source of the tremendous energy emitted by quasars and active galactic nuclei—supermassive black holes—will be discussed and the supporting evidence explained.

Progress Checklist

1. Quasars
- ❑ Quasars
- ❑ Redshifts and Distances
- ❑ Compact Energy Sources
- ❑ The Energy Problem
- ❑ Quasar Abundances
- ❑ Host Galaxies

2. Active Galactic Nuclei
- ❑ Active Galaxies
- ❑ Nonthermal Spectra
- ❑ Radio Galaxies
- ❑ Seyfert Galaxies
- ❑ BL-Lac Objects
- ❑ Comparisons to Quasars

3. A Unified Model of Quasars and AGNs
- ❑ Rotating Black Holes
- ❑ Evidence in M87
- ❑ Evidence in NGC4261
- ❑ Evidence in M84
- ❑ Evidence in Other Galaxies
- ❑ Feeding the Black Holes

Keywords

quasars
quasi-stellar objects (QSOs)
cosmological redshift
host galaxy
3C 273
active galaxies
active galactic nuclei (AGNs)
radio galaxies
BL-Lacertae objects (blazers)
Seyfert galaxies
nonthermal emission
synchrotron radiation
polarized light
optical jets
radio jets
supermassive black hole
Virial Theorem

Exercise 17-1: Introductory Narrative

Quasars were discovered in the early 1960s. The term *quasar* is a derivative of 1) ________ ____________________, so called because the first images of quasars were starlike in appearance. However, the spectra of quasars demonstrate extremely large 2) ________________ indicating that they have high recessional velocities. By applying 3) ________________ one can see that this implies that quasars are very distant objects. Since quasars can be seen at these large distances, they must be very luminous objects. Another interesting characteristic of quasars is that their brightness fluctuates over timescales of several days. Since the fastest possible velocity is the speed of light, quasars can only be several light-days in size. These observations encapsulate the enigma initially posed by quasars: they are about the size of a(n) 4) ____________________ but produce more energy than a typical 5) ______________.

Because almost all quasars are very distant, the 6) ____________________ times for them must also be very large. This implies that quasars are representative of the Universe of the past and that very few exist today. The distribution of quasars in space is evidence that the Universe 7) ____________________.

Active galaxies were originally thought to be unrelated to quasars. The 8) ____________ of these galaxies exhibit evidence of extremely energetic phenomena such as unusual structure and jets. Their spectra also typically contain 9) ____________________ radiation at longer wavelengths.

Our understanding of quasars and active galaxies has improved since the 1960s. Today, it is thought that active galaxies and quasars are very similar. Both are thought to be galaxies with very large 10) ____________________ at their centers and the major difference is the amount of material that is being devoured. Quasars are also more distant, and the stars of the surrounding galaxy are not as easily seen.

Exercise 17-2: True/False Questions

T / F 1. Quasars have very large redshifts, which implies (through Hubble's Law) that they must be at very great distances.

T / F 2. The fact that the luminosity of quasars varies over short timescales implies that they are extremely large objects.

T / F 3. The spectra of quasars exhibit very broad emission lines, further evidence that quasars are traveling away from us at high velocities.

T / F 4. That very few quasars have been found near us in space indicates that very few of them still exist today.

T / F 5. The observation that quasars were far more abundant in the past proves that the Universe evolves over time.

T / F 6. The spectra from active galaxies are dominated by blackbody radiation.

T / F 7. It is likely that the strong radio emissions detected from AGNs are caused by material shot out of compact nuclei by jets.

T / F 8. The major difference between active galaxies and quasars is that quasars are not part of a galaxy.

T / F 9. The existence of large black holes at the centers of active galaxies is supported by evidence for the existence of large amounts of unseen matter at the centers of galaxies.

T / F 10. Accretion disks in the cores of active galaxies can be identified by the redshift of material falling into the black hole.

T / F 11. Quasars are likely to "turn off" if they consume all of the available fuel in their galaxy.

T / F 12. "Turned off" quasars can "turn on" if new matter becomes available to spiral into their accretion disks due to interaction with another galaxy.

Unit 18
Cosmology

Chapter Objectives

Cosmology is the study of the Universe as a whole, as one single object—the ultimate "Big Picture." The Cosmological Principle, which states that on the large scale the Universe is homogeneous and isotropic, is the starting point for this branch of astronomy. Our large scale view of the Universe must be representative if we are to answer the big questions: what is the overall structure of the Universe? what is its history? its fate? In this chapter we will see how the expansion of the Universe (as evidenced by the galactic red shifts) and the application of Einstein's General Theory of Relativity lead naturally to a big bang beginning to our Universe. The possibilities for the ultimate fate of our Universe, whether the expansion will continue forever or rather be stopped and a resulting big crunch will occur, will be explored. The gravitational lensing that is a consequence of Einstein's view of gravity will be described and illustrated and its utility in finding the elusive dark matter that may seal our Universe's fate will be demonstrated.

The Hubble expansion of the Universe and the cosmic microwave background radiation are the two most important pieces of observable data concerning the Universe as a whole. In this chapter we will see how scientists apply these facts and the ideas of elementary particle physics to the first few seconds and minutes of our Universe's history. The few elementary particles (and antiparticles) that could exist in the extremely hot and dense early Universe will be introduced. The important changes that occurred in the early Universe as the temperature and density decreased will be explained. The inflationary versions of the big bang and the additional questions they answer will be discussed. The formation of large scale structure from a relatively smooth early universe in either a top-down or bottom-up scenario will be explored. The necessity for formulating a quantum version of gravity to understand the first small fraction of a second in our Universe's history will be described.

Progress Checklist / Cosmology

1. Cosmology
- ❑ Cosmological Issues
- ❑ Cosmology and Geometry
- ❑ Cosmological Principle
- ❑ Einstein Equations
- ❑ Cosmological Constant?
- ❑ Fate of the Universe

2. Gravitational Lensing
- ❑ Gravitational Lensing
- ❑ Lensing of Quasars
- ❑ Lensing of Galaxies
- ❑ Einstein Cross

3. Gamma Ray Bursts
- ❑ Gamma Ray Bursts
- ❑ Early Observations
- ❑ Local or Cosmological
- ❑ Other Wavelengths
- ❑ Redshifts and Distances
- ❑ Models

Progress Checklist / The Early Universe

1. The Big Bang

- ❑ The Big Bang
- ❑ Cast of Characters
- ❑ Equilibrium and Decoupling
- ❑ The First Three Minutes
- ❑ Subsequent Evolution
- ❑ Triumph of the Big Bang

2. Cosmic Background Radiation

- ❑ Discovery of the CBR
- ❑ Spectrum and Temperature
- ❑ Isotropy and Anisotropy
- ❑ Motion Relative to CBR
- ❑ Small Fluctuations
- ❑ Implications for Cosmology

3. Inflationary Universe

- ❑ Big Bang "Problems"
- ❑ Unification of the Forces
- ❑ Vacuum Energy
- ❑ Inflationary Expansion
- ❑ Solutions of the "Problems"
- ❑ Fluctuations and Structure

4. Formation and Structure

- ❑ Structure from Uniformity
- ❑ Top Down Theories
- ❑ Bottom Up Theories
- ❑ Role of Dark Matter
- ❑ Computer Simulations
- ❑ Where It Stands

5. The Planck Era

- ❑ The Planck Scale
- ❑ Quantum Gravitation
- ❑ Spacetime Foam
- ❑ Breakdown of Current Law?

Keywords / Cosmology

cosmology
expanding universe
big bang
hot big bang
balloon analogy
comoving coordinates
cosmological principle
Friedmann cosmologies
Einstein field equations
cosmological constant
vacuum energy density
cosmic inflation
the big crunch
open universe
closed universe
gravitational lensing
Einstein Cross
gamma ray bursts
BATSE
hypernova

Keywords / The Early Universe

big bang
radiation dominated universe
matter dominated universe
photon
proton
neutron
electron
positron (antielectron)
neutrino
antineutrino
antimatter
quarks
gluons
thermal equilibrium
decoupled
freeze-out
quark confinement
confinement transition
deuterium bottleneck
baryon
recombination transition
steady state model
perfect cosmological principle
cosmic background radiation (CBR)
blackbody curve
isotropy
anisotropy
horizon problem
dipole anisotropy
inflationary universe
flatness problem
magnetic monopole problem
Grand Unified Theories (GUTs)
elementary particle physics
lightcone
particle horizon
event horizon
strong interaction
electromagnetic force
weak force
gravitational force
Superunified Theories

Standard Model	inflationary epoch	Planck era
Standard ElectroWeak Theory	phase transition	quantum gravitation
spontaneous symmetry breaking	vacuum energy	superstring theory
Planck scale	cosmological constant	spacetime foam
	top-down theories	
	bottom-up theories	

Exercise 18-1: Introductory Narrative

Cosmology is the study of the overall structure and 1) ____________________ of the Universe. It attempts to answer the big questions like "How did the Universe come to be?" and "What will happen to it in the future?" Thus, cosmology is concerned only with distance scales of 2) ____________________ of galaxies and timescales of 3) ____________________ of years.

An observation central to cosmology is that all galaxies are traveling 4) ______________ from each other—the Universe is expanding. This suggests that the Universe originated in a high-density, high-temperature state. It then expanded to its present form in an ongoing explosion that is known as the 5) ____________________. The rate of expansion is slowing over time due to the mutual gravitational interaction of galaxies. Thus, the fate of the Universe (whether the expansion will continue forever) will be determined by the amount of matter in the Universe. The amount of matter necessary to stop the expansion is known as the 6) ____________________.

The fate of the Universe can also be discussed in terms of the geometry of space. If there is exactly enough matter to halt the expansion, space has no curvature; this is termed the 7) ____________________ universe. Insufficient matter to stop the expansion implies 8) ____________________ curvature and an open universe. Sufficient matter to halt the expansion implies positive curvature and a(n) 9) ____________________ universe. In this scenario the expansion becomes a contraction, and all the galaxies of the Universe would meet in a(n) 10) ____________________. Although all present methods of calculating the amount of matter in the Universe point to an open universe, the question is still unresolved due to the undetermined nature of 11) ____________________.

The Universe began in a state of very high temperature and density in a gigantic 12) ____________________ known as the big bang. Most of the energy of the early Universe was in the form of 13) ____________________. The simplicity of the early Universe can be seen through the small number of fundamental particles that existed: photons, protons, neutrons, electrons, neutrinos, and various 14) ____________________. Radiation and matter were in thermal equilibrium, and energy moved freely between the two forms in events of pair creation and pair 15) ____________________. As the temperature fell, the reaction rates for certain types of particles also fell, and eventually one by one the particles became 16) ____________________ from the general thermal equilibrium. It was still so hot that a typical photon had sufficient energy to break apart a neutron that had fused with a proton. This barrier to nuclear fusion is known as the deuterium 17) ____________________. Since free neutrons are not stable, their numbers were decreasing at this time. Once the temperature fell enough for deuterium to form, fusion of nuclei up to helium-4 occurred rapidly. This preserved the relative abundance of protons and neutrons at that time. Nuclear fusion was unable to build nuclei larger than helium-4 because there were no stable 18) ________________ or mass-8 isotope nuclei. After 1000 years, when the universe had cooled to a temperature of 100,000 K, most of the energy was in the form of 19) ________________.

The next major event in the big bang occurred at a time of 300,000 years when the Universe had cooled to a temperature of about 3000 K. It was now cool enough for electrons to remain bound to nuclei. Light no longer interacted with free electrons, and the Universe, which had been opaque, now became 20) ________________. We still detect the radiation released at this time, which is called the 21) ________________. The discovery of this radiation offered definitive proof to modern cosmologists that the big bang theory was correct. We are part of the big bang, an event that is still going on today as the Universe continues to expand and cool.

Exercise 18-2: True/False Questions

T / F 1. From any galaxy in the Universe, all other galaxies will be moving away from it with velocities given by Hubble's Law.

T / F 2. Determining the geometry of the Universe requires an understanding of the nature and abundance of dark matter.

T / F 3. If the Universe has positive curvature with enough mass to stop the present expansion, it is said to be open.

T / F 4. The Cosmological Principle states that on the scale of superclusters of galaxies the Universe is homogeneous and isotropic.

T / F 5. If the density parameter is less than one, then the Universe is headed toward a "big crunch."

T / F 6. Recent controversial evidence suggests that the expansion of the Universe is accelerating due to a nonzero vacuum energy density.

T / F 7. The gravitational lensing of quasars is strong evidence that they are really at the distances implied by their redshifts.

T / F 8. Efforts to identify MACHOs (a possible source for the missing baryonic mass) involve the MACHOs eclipsing the stars of the LMC.

T / F 9. Mapping the locations in the sky of gamma-ray bursts suggests that they are occurring in the disk of the Milky Way.

T / F 10. Gamma-ray bursts must release the radiation with very little interaction with matter, or they would be transformed into longer wavelength radiation.

T / F 11. Although there are several theories explaining gamma-ray bursts, all of them involve the collapse of spinning material to form a black hole.

T / F 12. Gamma-ray bursts can release almost as much energy as a nova.

T / F 13. The very early Universe was *matter dominated* because the energy was primarily in the form of nonrelativistic matter.

T / F 14. When a particle meets its antiparticle, they annihilate each other, meaning that they completely disappear with no trace.

T / F 15. The term *observable universe* is just another way of saying the entire universe.

T / F 16. An imbalance between the number of protons and neutrons occurred in the early Universe because free neutrons are less stable than protons and it was too hot for neutrons to form deuterium.

T / F 17. The nuclei of heavy elements such as carbon and nitrogen formed in the early Universe in a manner very similar to how they are formed in the cores of massive stars.

T / F 18. The relative abundances of deuterium and helium in the oldest stars serve as a good indicator of conditions in the early Universe.

T / F 19. The big bang theory predicts that there should be considerably more helium in the Universe than astronomers observe.

T / F 20. Since the time of the release of the microwave background radiation, its wavelength has increased and the number density of its constituent photons has decreased.

T / F 21. The "dipole anisotropy" of the COBE data is due to the Earth's orbital motion about the Sun.

T / F 22. It is difficult to explain how galaxies could have grown from the very small fluctuations in the CBR without involving more complex structure in dark matter.

T / F 23. The difficulty in explaining how parts of the CBR from opposite directions in the sky, which could never have been in contact, are at exactly the same temperature is called the horizon problem.

T / F 24. If a magnet is cut in half, north and south magnetic monopoles will be produced.

T / F 25. The top-down theory of galaxy formation predicts that masses the size of star clusters form first and later build galaxies through collisions and mergers.

T / F 26. The present observational evidence concerning the formation of structure favors a bottom-up assembly and an important role for cold dark matter.

T / F 27. Superstrings (which are theorized to be the fundamental building blocks of the Universe) are one-dimensional objects.

Unit 19
Life in the Universe

Chapter Objectives

It is a fascinating question: Are we alone? How do scientists consider the topic of life elsewhere in the Universe without dealing entirely in pure speculation or science fiction? The basic principles underlying the search for extraterrestrial life will be outlined in this chapter. The definition of life and its most fundamental biological components will be discussed. The probability of other solar systems with planets within their habitable zone (ecosphere) will be investigated. The possibility of communicating with an extraterrestrial intelligent civilization will be explored, and the criteria used to decide the proper method of communication will be discussed.

Progress Checklist

1. Biological Conditions
- ❑ Defining Life
- ❑ Origin of Life on Earth
- ❑ The DNA Code
- ❑ Timescales
- ❑ Evolution of Life
- ❑ Is Carbon Unique?

2. Astronomical Conditions
- ❑ Solar Systems
- ❑ Habitable Zones
- ❑ Timescales
- ❑ How Many Civilizations?

3. Communicating
- ❑ Distance and Time
- ❑ UFOs?
- ❑ Sending Messages
- ❑ Radio Searches
- ❑ SETI Projects
- ❑ What if They Reply?

Keywords

life	genes	habitable zone (ecosphere)
deoxyribonucleic acid (DNA)	chromosomes	Drake equation
cell	gene mutation	UFOs
adenine (A)	ribonucleic acid (RNA)	water hole
cytosine (C)	evolution	SETI
guanine (G)	natural selection	Project Phoenix
thymine (T)	carbon-based chemistry	

Exercise 19-1: Introductory Narrative

To accurately assess the likelihood of life existing on other planets, we must first make sure we understand the nature of life on Earth. Life began in the Earth's oceans about 1) ______________________________ years ago. Life is based upon the chemistry of the 2) ______________________________ atom. The building blocks of life (amino acids, etc.) were formed from the interaction of the Earth's primitive atmosphere, UV radiation, and lightning. The building blocks formed larger organic molecules through a process known as 3) ______________________________. At some point a molecule with the ability to reproduce itself was produced; that may be considered the first life. All living things carry information about their heritable traits in a long twisting molecule known as 4) ____________________. In successive generations of the species, these data can change through a process known as 5) ____________________. Some of these changes or 6) ____________________ are advantageous to the species due to a changing environment and help the species to survive in a process known as 7) ____________________.

Do the conditions that were necessary for the development of life on Earth exist on planets of other solar systems? Liquid water was certainly essential. Astronomers define the region around a star where liquid water can exist as the 8) ____________________, the size of which increases substantially for hotter stars. However, these hot stars live only a short time on the main sequence. Taking both of these factors into account suggests that spectral type 9) ____________________ stars are the most likely to have life evolve on planets orbiting them. One can attempt to estimate the number of intelligent civilizations in our galaxy using the 10) ____________________ equation.

Even if other intelligent civilizations have evolved in our galaxy, they are probably not near us in either space or time. Thus, sending messages as was done with the 11) ____________________ spacecraft in 1977 is one possible means of opening communication. A method more likely to be successful is sending and receiving radio signals. A signal was sent from the Arecibo Radio Observatory toward the globular cluster M13 in 1974. The NASA project SETI scanned the skies for signals, and its successor project 12) ______________ still does so today.

Exercise 19-2: Using the Drake Equation

The Drake equation is an attempt to calculate how many technically advanced civilizations might exist in our galaxy (or any other) with which we potentially could communicate. The number of civilizations is given by the product of the factors in the table to the right. The Drake equation is really more a model for thinking about the problem than a realistic method of making this calculation. Several of the factors in the Drake equation are extremely uncertain, and there is considerable debate among astronomers concerning their values.

$$N = N_S \cdot F_P \cdot N_H \cdot F_L \cdot F_I \cdot F_S$$

N is the number of communicative civilizations in the galaxy.

N_S is the number of stars in the galaxy.

F_P is the fraction of those stars with planets.

N_H is the average number of those planets in the habitable zone.

F_L is the fraction of those planets that evolve life.

F_I is the fraction of those for which life evolves to intelligence.

F_S is the fraction of the star's life in which the civilization is communicative.

The Drake Equation

The table below contains some consensus optimistic and pessimistic values for all of the parameters of the Drake equation and a column for you to enter your estimates. You should enter your estimates for each of the parameters in column 4, and state in column 5 whether you are closer to the optimists' or pessimists' estimate for this value; then multiply your column 4 values together to estimate the number of technically advanced civilizations in our galaxy. Your result will be just as valid as any astronomer's estimate.

Variable	Pessimistic Estimates	Optimistic Estimates	Your Estimates	Closer to Optimistic or Pessimistic
N_S	2×10^{11}	2×10^{11}		
F_P	0.01	0.5		
N_H	0.01	1		
F_L	0.01	1		
F_I	0.01	1		
F_S	10^{-8}	10^{-4}		
N	2×10^{-5}	1×10^{7}		

Are you an optimist or a pessimist? Which parameters do you think have the most uncertainty associated with them?

Exercise 19-3: True/False Questions

T / F 1. Life on the planet Earth is based on the complex chemistry of the element carbon.

T / F 2. Genes are combinations of chromosomes, which along with many proteins make up a single DNA molecule.

T / F 3. Damage from cosmic rays, solar UV, or natural radioactivity and alteration of DNA by chemicals in the environment all are possible sources of genetic mutation.

T / F 4. Laboratory experiments simulating the interaction of the Earth's primitive atmosphere and an energy source produce many of the chemical building blocks of life.

T / F 5. BillyBuzz the dragonfly might evolve larger wings in order to fly faster to escape the predatory birds that live near his pond.

T / F 6. Life is just as likely to develop on a planet in the habitable zone around a B2 spectral type star as around any other star.

T / F 7. Life is just as likely to develop on a planet orbiting a binary star system as on a planet orbiting any other star.

T / F 8. Life is just as likely to develop on a planet orbiting a star in a highly elliptical orbit as in any other scenario.

T / F 9. It is possible that Jupiter's moon Europa may have subsurface liquid oceans because Europa is within the Sun's habitable zone.

T / F 10. Spaceflight to other star systems is unlikely in the near future due to the extreme distances involved and the limitation that velocities must be less than the speed of light.

T / F 11. Radio waves with wavelengths between 18 and 21 cm are logical ways to send signals to extraterrestrials.

Appendix
Solutions to Exercises

Exercise 1-1: Introductory Narrative

1. celestial sphere 2. declination 3. east 4. proper motion 5. precession 6. diurnal motion 7. constellations 8. asterism 9. wanderer 10. Mars 11. moon 12. retrograde motion 13. ecliptic

Exercise 1-3: Following the Sun's Path for One Year

1. equator 2. northern 3. southern 4. summer solstice 5. northern 6. +23.5° 7. September 21 8. 0° 9. southward 10. southern 11. northern 12. December 21 13. winter solstice 14. southern 15. -23.5° 16. Arctic Circle

Exercise 1-5: Rising, Meridian, and Setting Times of Lunar Phases

1. 6 A.M. 2. 9 P.M. 3. First Quarter 4. Waning Crescent 5. 9 P.M. 6. 6 P.M. 7. Third Quarter 8. Waxing Crescent 9. Noon 10. 3 P.M. 11. Waxing Gibbous 12. First Quarter

Exercise 1-6: True/False Questions

1. **F**, The point directly above our heads is termed the zenith. The nadir is the point directly below us. 2. **F**, Right Ascension is analogous to longitude while declination is analogous to latitude. 3. **T** 4. **F**, On the summer solstice the sun is far north of the celestial equator. 5. **F**, Stars in a certain constellation need only be near each other on the celestial sphere and thus appear to be in the same direction in the sky from our perspective. They may still be at very different distances from us. 6. **F**, The Big Dipper is an asterism and part of the official constellation Ursa Major. 7. **T** 8. **F**, Planets normally move eastward. Retrograde is the occasional (apparent) westward motion. 9. **T** 10. **F**, The sidereal month is 27. 3 days long while the synodic month is 29.5 days long. 11. **F**, Its period of rotation is equal to its period of revolution about the Earth. To an observer out in space, the moon's rotation would be apparent. 12. **T** 13. **T** 14. **T**

Exercise 2-1: Introductory Narrative

1. Greek 2. crystalline 3. velocity 4. epicycles 5. heliocentric 6. Mars 7. ellipses 8. faster 9. square 10. cube 11. Copernican (or heliocentric) 12. telescope 13. craters on the moon (or sunspots) 14. Venus 15. Ptolemaic (or geocentric) 16. Jupiter 17. motion 18. Gravity 19. Kepler's 20. perturbations 21. general 22. gravitational

Exercise 2-2: Exploring Archaeoastronomy

This exercise is very open to interpretation and you should view the comments that follow as representative of what was in the mind of the author as opposed to what a student could reasonably be expected to theorize. This exercise is best explained in terms of a coordinate of the horizon coordinate system known as azimuth. This coordinate is an angle specifying a position relative to the observer on the horizon. It is 0° at the north point of the horizon and increases moving eastward where it is 90° at the east point, 180° at the south point, 270° at the west point, and just less than 360° back at the north point. The inner circle of stones is meant to indicate the position of the observer and the painted stones of the outer circle represent significant azimuth values. Thus, this pattern of stones is most similar to the Big Horn Medicine Wheel.

The icon directly north of the observer at an azimuth of 0∞ has a star with trails behind it. This could represent the north star Polaris which would be directly north of the observer at an altitude equal to the observer's latitude. Since Polaris is very near the North Celestial Pole it doesn't appear to rotate as other stars in the sky do. In fact it is always found at the exact same location and is very useful for navigation for people living in the northern hemisphere. Thus, all other stars trace out circular trails around Polaris.

The four icons with a sun symbol represent the extreme azimuth values of the rising and setting of the sun. The sun rises directly east (azimuth 90∞) on either equinox. It will rise north of there (ª60∞, the friendly sun) on the summer solstice which is the longest day of the year and south of there (ª120∞, the unfriendly sun) on the winter solstice.

The icons with crescent moons are very similar in nature representing the range of azimuth values for the rising and setting moon. The plane of the moon's orbit is inclined about 6∞ to the plane of the ecliptic. Thus the range of motion of the moon is slightly greater in azimuth than the range of motion of the sun.

The icons with the dog represent the rising and setting azimuth of the star Sirius which is the brightest star in the sky. Sirius is in the constellation Canis Major (the Big Dog) and it is commonly known as "the dog star." The star was especially significant to ancient people living near the Nile River. The heliacal rising of Sirius (when it first appears in the morning sky after a several month absence) was a sign that the river would soon flood.

Exercise 2-3: Applications of Kepler's Laws to Other Solar Systems

Simulation #1: Orbital Period of Sirius A and Sirius B = 50.8 years
Simulation #2: Separation of Kruger A and Kruger B = 9.99 AU

Exercise 2-4: Manifestations of Newton's Laws

1. Second Law 2. Law of Gravity 3. Third Law 4. First Law (or Second Law) 5. Law of Gravity 6. First Law (or Second Law) 7. Third Law 8. Second Law 9. Third Law 10. First Law (or Second Law)

Exercise 2-6: True/False Questions

1. **T** 2. **T** 3. **T** 4. **T** 5. **F**, The correct timeframe is around 200 B.C. 6. **F**, He used the shadow of the sun in wells at different locations. 7. **T** 8. **F**, Copernicus furthered the heliocentric theory that the sun was the center of the Solar System. 9. **T** 10. **F**, They couldn't observe parallax because it is extremely small for stars which are very far away. 11. **F**, Brahe is best known for making extremely accurate observations. 12. **T** 13. **T** 14. **F**, A circle has an eccentricity of 0 while an ellipse has an eccentricity that is greater than zero but less than 1. 15. **F**, The sun is located off center at one focus of the ellipse. 16. **T** 17. **T** 18. **F**, Galileo was the first person to use the telescope for astronomical purposes. Its invention is usually credited to Dutch spectacle maker Hans Lippershey. 19. **T** 20. **T** 21. **F**, They would both fall at the same rate in the absence of air friction. 22. **F**, The nuclei of atoms are held together by the strong nuclear force. 23. **F**, Your weight on the moon is about 1/6 of that on the Earth, but your mass is the same everywhere. 24. **T** 25. **T** 26. **T** 27. **F**, The shortening of objects in the direction of motion is known as length contraction. 28. **T** 29. **F**, The actual positions of distance objects are not changed, only their apparent position due to the bending of light. 30. **F**, Horizontal slicing would form a circle. 31. **F**, They are much too far away to exert much gravitational force. 32. **F**, It is only valid for systems that are not accelerating.

Exercise 3-1: Introductory Narrative

1. limitations (or deficiencies) 2. light gathering power 3. telescopes 4. efficiency 5. analog 6. CCDs 7. 70% 8. frequency (or wavelength) 9. atmospheric window 10. space based 11. twinkling 12. resolution

Exercise 3-2: Telescope Vocabulary

1. reflectors 2. mirrors 3. reflectors 4. refraction 5. chromatic aberration 6. convex 7. incidence 8. cassegrain 9. prime 10. Newtonian 11. square 12. resolution

Exercise 3-3: Powers of a Telescope

1. Yerkes 1-m refractor
   Light Gathering Power

$$\frac{(LGP)_A}{(LGP)_B} = \left(\frac{D_A}{D_B}\right)^2 = \left(\frac{100cm}{0.7cm}\right)^2 = 20{,}400$$

   Resolving Power would be limited by the atmosphere to 0.5″

2. Hubble Space Telescope
   Light Gathering Power

$$\frac{(LGP)_A}{(LGP)_B} = \left(\frac{D_A}{D_B}\right)^2 = \left(\frac{240cm}{0.7cm}\right)^2 \cong 118{,}000$$

   Resolving Power is around 0.1″ due to imperfections in the optics. However, Dawe's limit suggests the lower value

$$\alpha = \frac{11.6}{D} = \frac{11.6}{250} \cong 0.05 arc - \sec onds$$

Exercise 3-4: True/False Questions

1. **T** 2. **F**, Photographic plates only record about 1% of the light incident upon them, while CCDs recorded more than 70%. 3. **F**, The 21-cm line is produced by an electron flipping its spin. This is considerably lower energy light than that produced by an energy level transition. 4. **F**, Our atmosphere has windows in the visual and radio. 5. **T** 6. **F**, X-rays and Gamma rays are very energetic and indicate violent processes. 7. **T** 8. **T** 9. **T** 10. **T** 11. **F**, Cosmic Rays studies must be conducted from balloons and from orbit since they interact with atmospheric particles. 12. **T** 13. **F**, CCDs are much more efficient than either the eye or a photographic plate. 14. **T** 15. **T**

Exercise 4-1: Introductory Narrative

1. light 2. colors (or frequencies) 3. low 4. spectrograph 5. continuous 6. Planck Radiation 7. Wien's 8. Stefan-Boltzmann 9. emission 10. absorption 11. continuous 12. permitted (or discrete) 13. levels

Exercise 4-2: Energy, Wavelength and Frequency of E-M Waves

	Largest Wavelength	Largest Frequency	Largest Energy
1	C	A	A
2	C	B	B
3	A	C	C
4	C	B	B
5	A	B	B
6	B	A	A
7	C	B	B
8	C	A	A
9	B	A	A
10	A	C	C

Exercise 4-3: Energy Level Diagrams

IKMO emit(s) visible light; RTVX absorb(s) infrared light; ACEG emit(s) UV light; I emit(s) red light; N (or P) absorbs(s) violet light; QSUW emit(s) Paschen lines; IKMO emit(s) Balmer lines; ACEG emit(s) Lyman lines; K emit(s) H_β; J absorb(s) $H\alpha$; Q emit(s) the longest wavelength; G emit(s) largest frequency; H absorbs(s) shortest wavelength; R absorbs(s) smallest frequency; X represents ionization

Exercise 4-4 : Applying Radiation Laws to Stars

Star #1 Spica – T = 25,000K
Peak Wavelength = 1160 Angstroms
Energy Flux = 2.2×10^{13} ergs/s-cm^2
Ratio of Spica's Energy Flux compared to the Sun's Energy Flux = 345
Star #2 Betelgeuse – T = 3,000K
Peak Wavelength = 9670 Angstroms
Energy Flux = 4.6×10^{9} ergs/s-cm^2
Ratio of Betelgeuse's Energy Flux compared to the Sun's Energy Flux = 0.07

Exercise 4-5: True/False Questions

1. **T** 2. **F**, The spectral lines will be blueshifted. 3. **F**, The Zeeman Effect is due to magnetic fields. 4. **F**, Our atmosphere has "windows" in the visual and radio. 5. **T** 6. **F**, Typical stars would emit an absorption spectrum. 7. **T** 8. **F**, The intensity would decrease by a factor of 4. 9. **F**, Non-thermal radiation increases in intensity at long wavelengths. 10. **T** 11. **F**, Molecules absorb IR very strongly. 12. **T** 13. **F**, The dopper effect only tells you about radial velocity. Since our distance from the sun is relatively constant spectral lines do not show any appreciable dopper shift. 14. **T** 15. **T**

Exercise 5-1: Introductory Narrative

1. nine 2. plane 3. revolve 4. eccentricity 5. circular 6. large 7. density 8. nebular hypothesis 9. increased 10. gaps 11. radial velocity

Exercise 5-3: Simulating the Orbits of ExtraSolar Planets

Simulation #1:
Period = 1.21 years
Periastron Distance = 0.60 AU
Apastron Distance = 1.59 AU

Simulation #2: Period = 0.031 years

Question #1: Neither of these planets would fit in well with the planets of our Solar System. The massive Jovian planets of our Solar System are all found far from the Sun, while in both of these systems massive planets are found very close to the star. This is the case for the vast majority of planets that have been discovered orbiting other stars. Note that the planet orbiting HD 210277 has an orbit with very high eccentricity. The majority of planets in our Solar System have nearly circular orbits.
Question #2: These planets are all detected by the radial velocity or Doppler wobble technique. These wobbles are larger and easier to detect for massive planets that are in close orbits. Thus this technique is biased in favor of detecting this type of planet.

Recent theories trying to explain why very massive planets are orbiting so close to their stars have centered upon the interaction of massive planets that form far from the Sun. If several Jupiter-sized planets form nearby each other, it is possible that gravitational interactions would propel one of them into a tight and possibly highly eccentric orbit. Thus, most astronomers still think that massive planets form in the outer parts of Solar Systems even though at present we are not finding them there.

Exercise 5-4: True/False Questions

1. **F**, A planet's orbital velocity is only related to its distance from the Sun. Since Mercury is the closest planet to the Sun it has the largest orbital velocity 2. **F**, Venus is composed of rocky materials and has a high density while Saturn is a gas giant and has a low density. 3. **T** 4. **F**, Only a circular orbit has an eccentricity of zero. The values of all of the planets are fairly small, but all of them are ellipses that have eccentricities greater than zero. 5. **T** 6. **T** 7. **T** 8. **F**, Due to the conservation of angular momentum, the rate of rotation will speed up as it collapses. Remember that $L = mvr$, if r decreases then v must increase for L to retain the same value. 9. **T** 10. **T** 11. **T** 12. **F**, The majority of planets discovered around other stars today are massive planets found in the inner part of those Solar Systems. All of the massive planets in our Solar System orbit very far from the Sun. Many astronomers believe there is a statistical bias in using radial velocity techniques, in that they are very good at finding large planets that revolve in small orbits and thus, that is what they find.
13. **T** 14. **F**, Stars are most commonly found as part of a binary pair.

Exercise 6-1: Introductory Narrative

1. seismic 2. shear 3. differentiation 4. mantle 5. crust 6. dynamo effect 7. plate tectonics
8. carbon dioxide 9. oceans 10. nitrogen 11. oxygen
12. ozone 13. Coriolis 14. 1/4 15. 1/80 16. 5° 17. rotation 18. igneous 19. highlands
20. anorthosite 21. basalts 22. far (or dark) side 23. regolith 24. density 25. refractory
26. large impact

Exercise 6-2: True/False Questions

1. **T** 2. **F**, Shear waves cannot travel through the liquid outer core of the Earth 3. **T** 4. **F**, Earthquakes primarily along the boundaries where two crustal plates meet. 5. **T** 6. **F**, Our atmosphere is referred to as an oxidizing atmosphere due to the large amounts of oxygen present. 7. **F**, The ozone primarily absorbs UV radiation from the sun. 8. **T** 9. **T** 10. **F**, The projectile would be deflected eastward. 11. **F**, The northern lights are caused by high energy particles from the sun that have been trapped by the Earth's magnetic field colliding with molecules in the Earth's atmosphere. 12. **T** 13. **T** 14. **T** 15. **F**, The seismic activity on the moon is minimal and is due to tidal forces from the Earth. 16. **T** 17. **F**, Although sedimentary rocks are the most common type found on the surface of the Earth, on the moon igneous rocks are the most common. 18. **T** 19. **F**, Most of the moons in our Solar System keep one face toward their parent planet. 20. **T** 21. **F**, There were many more collisions in the early Solar System since there was considerable debris left over from planet formation. 22. **T** 23. **F**, Lunar volcanoes tend to surround maria indicating their origin in meteor impacts. 24. **F**, The moon has been geologically dead for over 3 billion years. 25. **T** 26. **T** 27. **T**

Exercise 7-1: Introductory Narrative

1. 0.4 AU 2. greatest elongation 3. velocity 4. 59 5. magnetic field 6. cratering 7. heavy metals 8. 0.7 AU 9. retrograde 10. greenhouse 11. 90 12. sulfuric acid 13. Magellan 14. seasons 15. density 16. Red Planet 17. dust storms 18. Olympus Mons 19. plate tectonics 20. canals 21. thicker (more dense) 22. Viking

Exercise 7-2: The Rotation of Mercury and Venus

Time for one rotation = 59 days
Time for one revolution = 88 days
Time for 3 rotations = approximately 176 days
Time for 2 revolutions = approximately 176 days

Question #1: The ratio of the rotational period to the orbital period is 2:3, and is an example of more complicated tidal locking than for the Earth-Moon system.

Question #2: Venus rotates at a much slower rate and in the opposite direction that it revolves which is called retrograde rotation.

Exercise 7-3: True/False Questions

1. **F**, The two densities are very comparable. 2. **T** 3. **T** 4. **F**, Hotter temperatures are found on Venus due to the greenhouse effect. 5. **F**, Mercury has an abundance of heavy metals. It is thought that it has lost some of its lighter density crust and mantle material from impacts. 6. **T** 7. **T** 8. **T** 9. **F**, The atmosphere of Venus is mostly carbon dioxide. There are no oceans to absorb carbon dioxide as on the Earth. 10. **T** 11. **F**, Mars actually has a substantially lower density than those of Mercury, Venus, or the Earth. 12. **T** 13. **T** 14. **F**, The Martian crust must be very thick to support the weight of such a monstrous volcano. 15. **F**, The orbital tilt of Mars is very similar to that of the Earth and thus it would have seasons with comparable variations in weather. 16. **T** 17. **F**, The positive results are now thought to be the result of complex chemistry in the Martian soil but not indicative of life. 18. **F**, Phobos rises in the west and sets in the east because its orbital velocity is extremely large in that direction. 19. **T** 20. **T** 21. **T**

Exercise 8-1: Introductory Narrative

1. 318 2. density 3. belts 4. Great Red Spot 5. solid surface 6. liquid metallic hydrogen 7. magnetic field 8. Galileo 9. Io 10. sulfur 11. magnetosphere 12. water 13. Ganymede 14. cratered 15. asteroids 16. density 17. helium 18. anticyclonic cells 19. winds 20. ringlets 21. Cassini's Division 22. ice crystals 23. spokes 24. shepherd moons 25. 18 26. atmosphere 27. methane 28. hydrocarbons 29. Cassini 30. Jovian 31. density 32. elliptical (or eccentric) 33. ecliptic 34. seasons 35. perpendicular 36. collision 37. velocity 38. Great Dark Spot 39. Pluto 40. Inclination 41. Charcn

Exercise 8-2: Characteristics of Terrestrial and Jovian Planets

1. Jovian 2. Both 3. Jovian 4. Terrestrial 5. Terrestrial 6. Both 7. Both 8. Terrestrial 9. Jovian 10. Both 11. Both 12. Jovian 13. Neither 14. Jovian 15. Terrestrial 16. Jovian 17. Terrestrial 18. Neither 19. Jovian 20. Jovian

Exercise 8-3: True/False Questions

1. **T** 2. **T** 3. **F**, The rapid rotation causes Jupiter to be fatter at the equator, which is known as oblateness. 4. **T** 5. **F**, Jupiter's powerful magnetosphere would deflect any charged particles from the solar wind. 6. **T** 7. **T** 8. **F**, Callisto is a dead world. Io has the active volcanoes. 9. **T** 10. **F**, It cooled very quickly, which didn't allow enough time for differentiation. 11. **F**, The small size and irregular shapes suggest that they are likely to be captured asteroids. This has little to do with their density, although most asteroids have fairly high densities. 12. **T** 13. **T** 14. **F**, Saturn has a lower temperature than Jupiter which causes the chemical reactions which color the upper atmosphere to proceed at a slower rate. 15. **F**, It is now known that all of the Jovian planets have ring systems. 16. **T** 17. **F**, Chemical reactions typically proceed faster where it is warmer. 18. **T** 19. **F**, Objects inside the Roche Limit will only be "pulled apart" is they are held together solely by gravity. 20. **T** 21. **F**, The fact that the rings are bright and not covered by dust and the difficulty involved in maintaining stable orbits for long periods of time both suggest that the rings are younger than Saturn. 22. **T** 23. **F**, Titan is much too cold for liquid water. However, methane and ethane are still liquid at Titan's temperature. 24. **F**, Jupiter has no medium sized moons comparable to Saturn's six icy moons. 25. **T** 26. **T** 27. **T** 28. **F**, Uranus has no heat source. Neptune has a substantial heat source which explains its energetic weather patterns. 29. **T** 30. **F**, Actually their albedos (percentage of incident light reflected) are very low due to their being covered with dark organic compounds. 31. **F**, The rings were discovered before the time of Voyager by stellar occultation. 32. **T** 33. **T** 34. **T** 35. **T** 36. **F**, Pluto's escape velocity is very small because its mass is small. It can retain an atmosphere because it is so cold that gas molecules move very slowly. 37. **F**, Pluto was inside the orbit of Neptune from 1979 to 1999. It has since moved back outside and is the farthest planet from the sun. 38. **F**, This was a commonly cited theory in the past but has recently fallen out of favor.

Exercise 9-1: Introductory Narrative

1. dirty snowballs 2. short period 3. Kuiper Belt 4. Oort Cloud 5. nucleus 6. coma 7. gas (or plasma) 8. dust 9. 76 10. Hale-Bopp 11. Asteroid Belt 12. 500,000 13. Jupiter 14. Kirkwood Gaps 15. collision 16. 1500 17. 4 18. dinosaur 19. meteor 20. meteorite 21. meteor shower 22. radiant 23. comet

Exercise 9-2: True/False Questions

1. **T** 2. **T** 3. **F**, Long period comets likely originate in the Oort Cloud. 4. **F**, A comet nucleus is only a few kilometers in diameter. 5. **F**, Comet densities are typically about 0.25 g/cc while terrestrial planets have densities around 5 g/cc. 6. **F**, It is the gas (or plasma) tail that always points away from the sun. 7. **F**, Halley's Comet has an elliptical orbit going out beyond the orbit of Pluto. 8. **T** 9. **T** 10. **T** 11. **T** 12. **F**, In non-linear systems the response is not proportional to the stimulus. 13. **F**, Asteroids are more likely material left over from the formation of our Solar System that never formed a planet. 14. **T** 15. **F**, The radiant is the point in the sky from where the majority of meteors appear to be coming. 16. **T** 17. **T** 18. **F**, The most common "falls" are stony meteorites. The most common "finds" are iron meteorites due to their distinctive appearance and ability to withstand weathering. 19. **F**, The number of meteorites that fall there is no larger than anywhere else on earth. One factor is that there are no other rocks on the ice from which one would have to differentiate a meteorite. 20. **T** 21. **T** 22. **T** 23. **F**, This conclusion was made by a small group of researchers and is highly disputed by others. There is no consensus among researchers on whether the Martian meteorites really show evidence of Martian life.

Exercise 10-1: Introductory Narrative

1. helium 2. core 3. convection 4. hydrostatic equilibrium 5. luminosity 6. photosphere 7. granulation 8. chromosphere 9. corona 10. magnetic 11. Sunspots 12. flares 13. solar wind

Exercise 10-2: True/False Questions

1. **T** 2. **T** 3. **F**, The solar constant is only the amount of energy passing through 1 square meter of the surface of that sphere. 4. **F**, Radiative transport is the most important mechanism in the inner parts of the Sun. Convection only becomes significant near the Sun's surface. 5. **T** 6. **F**, The edge of the Sun will be darker, a phenomenon known as limb darkening. 7. **T** 8. **T** 9. **F**, The number of sunspots varies with an 11-year periodicity. 10. **T** 11. **F**, The Babcock model states that sunspots occur due to convection being inhibited. 12. **T** 13. **T**

Exercise 11-1: Introductory Narrative (Stellar Energy Production)

1. energy 2. nuclear (or fusion) 3. stable 4. positively 5. Coulomb barrier 6. temperature 7. helium 8. proton-proton 9. convection 10. massive 11. half

Exercise 11-2: Introductory Narrative (Stellar Parameters)

1. parallax 2. Hipparchos 3. proper motion 4. Doppler shift 5. brightness 6. distance 7. absolute magnitude 8. temperature 9. B 10. K 11. color index 12. Hertzsprung-Russell 13. radius

Exercise 11-3: Introductory Narrative (Binary Stars and Clusters)

1. visual 2. astrometric 3. spectroscopic 4. radial velocity curves 5. eclipsing 6. 90° 7. light curve 8. radii 9. accretion disk 10. Open 11. Globular 12. Population II

Exercise 11-4: Using Parallax

3. The length is approximately 33.5 blocks. 4. $(33.5\ blocks)\left(\frac{20\ m}{1\ block}\right) = 670\ m$

Exercise 11-5: Using Magnitudes

1. α (smallest m) 2. κ (largest m) 3. ι ($16 \Rightarrow m_2 - m_1 = 3$) 4. β or γ ($100 \Rightarrow m_2 - m_1 = 5$) 5. 16 ($m_2 - m_1 = 6 - 3 = 3$) 6. 40 ($m_2 - m_1 = 5 - 1 = 4$) 7. 4 (at 10 pc, $m = M$) 8. –2 ($m - M = 3 \Rightarrow 1 - M = 3$) 9. 160 [$m - M = 2 - (-4) = 6$] 10. 16 [$m - M = 1 - 0 = 1$] 11. –1.5 (140 pc $\Rightarrow m - M = 5.5$) 12. 9 [$m - M = 2 - (-7) = 9$]

Exercise 11-6: The HR Diagram

1. Spica 2. Betelgeuse 3. Rigel 4. Procyon B 5. Betelgeuse—Large stars are found to the upper right of the HR Diagram. 6. Sirius B or Procyon B (It should be noted that some astronomers don't refer to white dwarfs as stars because they no longer produce energy through nuclear fusion reactions. They are referred to as compact objects—objects that used to be stars.) 7. Spica, Altair, Procyon A, and the Sun (Vega and Sirius are very close) 8. Vega, Sirius A, or Deneb 9. Rigel, Deneb, or Betelgeuse 10. Betelgeuse—A star must be very cool to have molecular lines in its spectrum.

Exercise 11-7: Spectroscopic Parallax

Star 1: Spectral Type = B5
Luminosity Class = V
Absolute Magnitude = –2
Distance Modulus = 11
Distance (in pc) = 1600
Star 2: Spectral Type = B5
Luminosity Class = Ia or Ib
Absolute Magnitude = –7
Distance Modulus = 11
Distance (in pc) = 1600
Star 3: Spectral Type = K2
Luminosity Class = II
Absolute Magnitude = –2
Distance Modulus = 8
Distance (in pc) = 400
Star 4: Spectral Type = O8
Luminosity Class = V
Absolute Magnitude = –5
Distance Modulus = 12
Distance (in pc) = 2500
Star 5: Spectral Type = K2
Luminosity Class = III
Absolute Magnitude = 0
Distance Modulus = 10
Distance (in pc) = 1000
Star 6: Spectral Type = G0
Luminosity Class = IV
Absolute Magnitude = 2
Distance Modulus = 6
Distance (in pc) = 160

Exercise 11-8: Eclipsing Binary Light Curves

1. To simplify matters, we will make an assumption concerning the orbital velocities of the stars. Assume that star A is moving out of the page toward the observer in frame I. Answers are given in the first orbital cycle.
   Frame I—5 days
   Frame II—This could be either 10 days or 33 days.
   Frame III—14 days
   Frame IV—16 days
   Frame V—This could be either 20 days or 25 days.
   Frame VI—29 days
2. Approximately 33 days
3. Primary eclipse—1.0 magnitude; secondary eclipse—0.4 magnitude

Exercise 11-9: Spectroscopic Binary Stars

1. Frame I—5 days
   Frame II—20 days
   Frame III—13 days
   Frame IV—0 days
2. Approximately 26 days
3. –15 km/s
4. Star B is the more massive because its radial velocity has a smaller range of values.

Exercise 11-10: The Ages of Open Clusters

1. Upper Left Cluster—Turnoff point is approximately 9750 K.
   —Age is approximately 400 My.
   —This cluster contains two blue stragglers.
2. Upper Right Cluster—Turnoff point is approximately 14,000 K.
   —Age is approximately 100 My.
3. Lower Left Cluster—Turnoff point is approximately 16,000 K.
   —Age is approximately 75 My.
4. Lower Left Cluster—Turnoff point is approximately 6,500 K.
   —Age is approximately 6,000 My.
   —This cluster contains two blue stragglers.

Exercise 11-11: True/False Questions

1. **F,** The energy from gravitational contraction is very small. 2. **F,** A fusion reaction involves the combining of two small nuclei. 3. **T** 4. **T** 5. **F,** The CNO Cycle is important in massive stars. 6. **T** 7. **T** 8. **F,** Neutrinos interact very weakly with matter and often escape from stars without interacting with matter at all. 9. **T** 10. **F,** The space-based Hipparchos satellite measured parallax to an accuracy of a thousandth of an arc second. 11. **T** 12. **F,** The V filter is very similar to the eye, while the B filter is most similar to photographic film. 13. **T** 14. **F,** Different spectral types are predominantly due to stars having different surface temperatures. Composition is a minor influence. 15. **T** 16. **F,** Both stars orbit around the center of mass. Although Sirius A will be closer to this point than Sirius B, it certainly is not stationary. 17. **F,** This relation is only valid for main sequence stars. It doesn't apply to red giants or to white dwarfs. 18. **F,** One can only determine the total mass of the pair of stars. 19. **T** 20. **T** 21. **F,** When the separation between source and observer is increasing, spectral lines will be redshifted. 22. **F,** For eclipses to occur, the angle of inclination i must be near 90°. 23. **F,** The B8 star has a much higher surface temperature than the K2 star; thus, the deeper eclipse must occur when the K2 star eclipses it. 24. **T** 25. **F,** Novae involve the accretion of material onto the surface of a white dwarf. 26. **T** 27. **T** 28. **T** 29. **F,** Metals are formed inside stars over time. The stars in globular clusters are very old but contain very small amounts of metals because not many metals had been made yet at the time the stars formed. 30. **T**

Exercise 12-1: Introductory Narrative

1. molecular clouds 2. Jeans 3. shock waves 4. fragmentation 5. energy 6. energy transport 7. cocoon 8. bipolar flows 9. 30 million 10. evolutionary track 11. hydrogen 12. 0.8

Exercise 12-2: Luminosities and Lifetimes of Main Sequence Stars

Star #1: Proxima Centauri $M = 0.1M_\odot$

$$L = (M)^{3.5} = (0.1M_\odot)^{3.5} = 3.2 \times 10^{-4} L_\odot$$

$$T = \frac{1}{M^{2.5}} = \frac{1}{(0.1M_\odot)^{2.5}} = 316T_\odot$$

$$316T_\odot = 316T_\odot \left(\frac{1 \times 10^{10}\ years}{1T_\odot} \right) = 3.12 \times 10^{12}\ years$$

Star #2: Rigel $M = 10\ M_\odot$

$$L = (M)^{3.5} = (10M_\odot)^{3.5} = 3162L_\odot$$

$$T = \frac{1}{M^{2.5}} = \frac{1}{(10M_\odot)^{2.5}} = 316 \times 10^{-3} T_\odot$$

$$3.16 \times 10^{-3} T_\odot = 3.16 \times 10^{-3} T_\odot \left(\frac{1 \times 10^{10}\ years}{1T_\odot} \right) = 3.16 \times 10^{7}\ years$$

Exercise 12-3: True/False Questions

1. **T** 2. **F,** The temperature will rise. 3. **F,** Hydrostatic equilibrium refers to a state of balance between expansion due to pressure and contraction due to gravity. A star is not in hydrostatic equilibrium if it is expanding or contracting. 4. **T** 5. **T** 6. **T** 7. **F,** UV radiation merely evaporates the low-density material surrounding the EGGs so that they may be seen. 8. **T** 9. **F,** Collapsing protostars are fully convective. Energy cannot flow efficiently by radiative transport. 10. **F,** Massive stars will collapse much more quickly than low-mass stars. 11. **T** 12. **F,** The width is due to differing compositions, which affect opacities. 13. **T** 14. **F,** Although lithium is useful, it is because it is destroyed in stellar fusion reactions. Thus, if lithium is present, it is assumed that the object has never had nuclear reactions. 15. **T**

Exercise 13-1: Introductory Narrative

1. 10 billion 2. upper right 3. hydrogen 4. red giant 5. helium flash 6. horizontal branch 7. red giant 8. planetary nebula 9. left 10. electron pressure 11. white dwarf 12. Earth

Exercise 13-2: Using Pulsating Variable Stars as Distance Indicators

Star #1: $m - M = 9.5 - 0.5 = 9$

This corresponds to a distance of 630 pc.

Star #2: From the graph one can estimate that $M = -1$.

$m - M = 4 - (-1) = 5$

This corresponds to a distance of 100 pc.

Star #3: From the graph one can estimate that $M = -4$.

$m - M = 6 - (-4) = 10$

This corresponds to a distance of 1000 pc.

Exercise 13-3: True/False Questions

1. **T** 2. **T** 3. **F,** The Sun will first burn hydrogen in a shell surrounding the helium ash core. It is necessary for the helium core to collapse and get much hotter before helium fusion can begin. 4. **T** 5. **T** 6. **F,** Planetary nebulae really have nothing to do with planets. 7. **F,** There are no fusion reactions occurring in a white dwarf. 8. **F,** The 1.4–1.5 solar mass value is an upper limit for the mass of a white dwarf. 9. **T** 10. **F,** Although Cepheids are useful as distance indicators because of their period-luminosity relation, the luminosity gets larger for stars with larger pulsation periods. 11. **F,** Type I supernovae involve accretion onto a white dwarf in a binary system. There are no hydrogen lines because there is very little hydrogen in a white dwarf. 12. **T** 13. **T** 14. **F,** The most stable nuclei are those in the "iron peak" with mass numbers between 55 and 60. 15. **F,** The r-process requires an abundance of heavy nuclei like iron and a strong neutron source. These conditions are likely to be present in a Type II supernova.

Exercise 14-1: Introductory Narrative

1. white dwarfs 2. 1.44 solar masses 3. protons 4. a city 5. magnetic 6. rapidly 7. synchrotron 8. pulsar 9. lighthouse 10. singularity 11. event horizon 12. charge 13. X-rays

Exercise 14-2: True/False Questions

1. **T** 2. **F,** The lower limit for the mass of a neutron star is 1.44 solar masses. Stellar remnants with lower masses than that can become white dwarfs. 3. **F,** They are likely the imploded cores of Type II supernovae, which are formed from massive stars. Type I supernovae involve white dwarfs in binary systems. 4. **T** 5. **T** 6. **F,** A nova occurs when matter accretes onto a white dwarf in a binary system. Although the phenomenon involving a neutron star is very similar, it is known as an X-ray burster. 7. **T** 8. **F,** The two axes generally are not aligned. 9. **F,** The rate of spinning slows as a pulsar get older because it is slowly radiating away energy. 10. **T** 11. **F,** Millisecond pulsars spin so rapidly that they must have gained angular momentum from some external source—probably a binary companion. 12. **T** 13. **T** 14. **F,** The statement refers to the fact that all information about the matter that makes up a black hole is lost. Black holes have very few distinguishing characteristics. 15. **T**

Exercise 15-1: Introductory Narrative

1. 200,000,000,000 2. Harlow Shapley 3. Edwin Hubble 4. central bulge 5. slowly 6. low 7. II 8. rapidly 9. I 10. rotation curve 11. dark matter

Exercise 15-2: The Components of the Milky Way Galaxy

1. D 2. H 3. D 4. H 5. D 6. D 7. H 8. H 9. H 10. D 11. H 12. H 13. D 14. D 15. H

Exercise 15-3: True/False Questions

1. **T** 2. **F,** Galactic latitude is being described. 3. **F,** All stars produce more visual light than infrared. The reason we can see the central region of the Milky Way better in infrared is that it isn't obscured as much by the intervening dust. 4. **F,** A population I star is high in metals that are produced by nucleosynthesis over time. Thus, a population I star is likely to be much younger. 5. **T** 6. **F,** Disk stars have circular coplanar orbits. Halo stars have randomly oriented elliptical orbits. 7. **T** 8. **F,** The spiral arms are regions of active star formation, and they appear bright because of the hot massive stars that form there. 9. **T** 10. **T** 11. **F,** Rotation curves definitely suggest some type of dark yet gravitational matter, but there are many other possible explanations for this than black holes. 12. **T** 13. **F,** HII regions have hydrogen atoms that have been ionized. HII regions are much hotter than clouds with molecular hydrogen. 14. F Interstellar reddening is due to the dust grains of the interstellar medium, which are much larger than gas molecules. 15. **T**

Exercise 16-1: Introductory Narrative

1. irregular 2. bars 3. spherical (round) 4. large 5. loosely 6. Hubble Tuning Fork Diagram 7. dust 8. star formation 9. evolve 10. superclusters 11. gravitational 12. dark matter

Exercise 16-2: Distances to Galaxies

Galaxy #1: $m - M = 20.5 - (-17.5) = 38 \rightarrow d = 4 \times 10^8$ pc = 400 Mpc
Galaxy #2: $m - M = 14.3 - (-19.7) = 34 \rightarrow d = 6.3 \times 10^7$ pc = 63 Mpc

Exercise 16-4: True/False Questions

1. **F,** The Hubble Tuning Fork Diagram merely classifies galaxies on the basis of structure. It says nothing about galaxy evolution. 2. **T** 3. **F,** Spiral galaxies have far more gas and dust than do elliptical galaxies. 4. **F,** Spiral arms are visible because they are regions of active star formation and have massive, extremely luminous stars in them. These star form there and don't live very long. Thus, they are seen only in the spiral arms because they die before they can move out. 5. **F,** There is a substantial observational bias in favor of spiral galaxies because they are on average much brighter than elliptical galaxies. Thus, we see many more spiral galaxies even though elliptical galaxies are considerably more abundant. 6. **T** 7. **T** 8. **F,** A mass to light ratio of 400 implies the presence of considerable material that is not producing light—dark matter. 9. **T** 10. **F,** The motion toward the Great Attractor is a small local (peculiar) effect superimposed on the expansion of the Universe known as the Hubble flow. 11. **T** 12. **T** 13. **T** 14. **F,** A typical mature spiral galaxy like the Milky Way produces 2 or 3 new stars per year. In a starburst galaxy the rate is about 100 times larger.

Exercise 17-1: Introductory Narrative

1. quasi-stellar 2. redshifts 3. Hubble's Law 4. solar system 5. galaxy 6. look-back 7. evolves 8. cores (centers) 9. nonthermal 10. black holes

Exercise 17-2: True/False Questions

1. **T** 2. **F,** For a quasar's luminosity to change over a short period of time, it must be a small object because whatever physical process is causing the change in luminosity must travel slower than the speed of light. 3. **F,** Broad lines are caused by random velocities. 4. **T** 5. **T** 6. **F,** Most of the radiation coming from active galaxies is nonthermal synchrotron radiation. 7. **T** 8. **F,** Quasars are thought to be the superluminous centers of distant galaxies; they are so far away that the surrounding galaxy cannot be seen as it can for nearby active galaxies. 9. **T** 10. **F,** The radial velocity near the cores of active galaxies shows an "S pattern" (both a redshift and a blueshift) from the material rotating in the accretion disk. 11. **T** 12. **T**

Exercise 18-1: Introductory Narrative

1. origin 2. superclusters 3. billions 4. away 5. big bang 6. critical density 7. flat 8. negative 9. open 10. big crunch 11. dark matter 12. explosion 13. radiation 14. antiparticles 15. annihilation 16. decoupled 17. bottleneck 18. mass-5 19. matter 20. transparent 21. cosmic background radiation

Exercise 18-2: True/False Questions

1. **T,** (Ignoring any peculiar motions of cluster members) 2. **T** 3. **F,** What is described is a closed universe. 4. **T** 5. **F,** For the Universe to collapse back on itself, the density parameter must be greater than one. 6. **T** 7. **T** 8. **F,** MACHOs could be detected by the lensing (short term brightening) of the light from distant stars. 9. **F,** The distribution of gamma-ray bursts is isotropic, and redshift observations indicate that they are not limited to our galaxy. 10. **T** 11. **T** 12. **F,** Gamma-ray bursts release as much or more energy as a supernova. 13. **F,** The early Universe was radiation dominated. 14. **F,** The annihilation must still satisfy the conservation of mass-energy; thus, photons with equal energy are produced. 15. **F,** The observable universe (the part we can observe) is a subset of the entire universe. 16. **T** 17. **F,** Nuclei more massive than helium-4 could not form in the early Universe in any appreciable amount. 18. **T** 19. **F,** There is good agreement between estimates of helium abundance from cosmology and those from observation. 20. **T** 21. **F,** The dipole anisotropy is due to the motion of the Local Group of galaxies, which move at about 600 km/s relative to the CBR. 22. **T** 23. **T** 24. **F,** One simply gets two smaller magnets. It is not possible to form a magnetic monopole today. 25. **F,** What is described is the bottom-up theory. 26. **T** 27. **T**

Exercise 19-1: Introductory Narrative

1. 3.0–3.6 billion 2. carbon 3. chemical evolution 4. DNA 5. evolution 6. mutations 7. natural selection 8. habitable zone 9. F-G-K 10. Drake 11. Voyager 12. Phoenix

Exercise 19-3: True/False Questions

1. **T** 2. **F,** Actually, chromosomes are combinations of genes that make up DNA. 3. **T** 4. **T** 5. **F,** Evolution is a very time-consuming process. BillyBuzz cannot evolve in any way. However, if he happens to have larger wings than other dragonflies, he is more likely to escape predatory birds and pass that characteristic down to his offspring. 6. **F,** A B2 star has a very short main sequence lifetime. 7. **F,** There are only a small number of stable orbits for planets orbiting binary stars. 8. **F,** An elliptical orbit would cause the temperature of a planet to fluctuate between extremes. 9. **F,** Europa is far outside the habitable zone. If Europa has subsurface liquid water, it is due to tidal heating by Jupiter. 10. **T** 11. **T**